Glen Wright is an academic. Sort of. Actually, he started his PhD in 2012 and is yet to finish. In the meantime, he started Academia Obscura, a blog about the lighter side of academic life. Born in the Black Country, Glen now lives in Paris, where he works for a non-governmental organisation trying to save the ocean. Neither is as glamorous as it sounds. Glen finds writing about himself in the third person extremely uncomfortable, but is equally uncomfortable breaking with accepted convention in his first book.

With special thanks
to Problem Solvers,
Literature and Latte

literatureandlatte.com/scrivener

ACADEMIA OBSCURA

The hidden silly side
of higher education

GLEN WRIGHT

Unbound

First published by Unbound in 2017
This edition published in 2018

Unbound
6th Floor Mutual House, 70 Conduit Street, London W1S 2GF

www.unbound.com

Text Design by carrdesignstudio.com

A CIP record for this book is available from the British Library

ISBN 978-1-78352-694-9 (trade pbk)
ISBN 978-1-78352-341-2 (trade hbk)
ISBN 978-1-78352-342-9 (ebook)
ISBN 978-1-78352-343-6 (limited edition)

Printed in Great Britain by CPI Group (UK)

To my late PhD

With special thanks to
Mice Chancellor Palimpsest Parnassus

Dear Reader,

The book you are holding came about in a rather different way to most others. It was funded directly by readers through a new website: Unbound. Unbound is the creation of three writers. We started the company because we believed there had to be a better deal for both writers and readers. On the Unbound website, authors share the ideas for the books they want to write directly with readers. If enough of you support the book by pledging for it in advance, we produce a beautifully bound special subscribers' edition and distribute a regular edition and ebook wherever books are sold, in shops and online.

This new way of publishing is actually a very old idea (Samuel Johnson funded his dictionary this way). We're just using the internet to build each writer a network of patrons. At the back of this book, you'll find the names of all the people who made it happen.

Publishing in this way means readers are no longer just passive consumers of the books they buy, and authors are free to write the books they really want. They get a much fairer return too – half the profits their books generate, rather than a tiny percentage of the cover price.

If you're not yet a subscriber, we hope that you'll want to join our publishing revolution and have your name listed in one of our books in the future. To get you started, here is a £5 discount on your first pledge. Just visit unbound.com, make your pledge and type **academia5** in the promo code box when you check out.

Thank you for your support,

Dan, Justin and John
Founders, Unbound

ACKNOWLEDGEMENTS

I would like to thank the wonderful people who provided love, support, inspiration, and proofreading. Without them, I'd probably never have started writing this book, much less finish it.

- Bart Wasiak, for the bet that led to this book, the support and encouragement during its development, and for going above and beyond with a last-minute edit.

- Emily Gong♥ for giving me unfettered access to her apparently bottomless well of love and support, and for putting up with far more of my nonsense than most.

- Jill Cooper, the biggest individual donor to the crowdfunding effort. Thanks Mom!

- Haydn Griffith-Jones for being a great friend, even if he never did get around to reading my drafts (see page 100, mate).

- Johannes Krebs, fellow PhDiva and the best writing buddy one could ask for.

- Julien Rochette, a fantastic boss and mentor. Merci chef pour m'avoir appris comment faire mousser, jouer du pipeau, et danser la gigoulette.

- Harriet Harden-Davies, whose fascinating science history lessons, boundless enthusiasm, and love of the ocean and strong dark beer, have unwittingly plunged her into my inner circle.

The following people have talked me down from various stages of deadline-induced panic and impostor syndrome, and done a whole heap of proofing and editing:

- Josh Bernoff (withoutbullshit.com), author of the excellent *Writing Without Bullshit*, who helped clean up my rambling and made me realise how much passive voice was being used.

- Katrin Boniface, for the grammar pointers.

- Stevyn Colgan, for reassurance, support, and last-minute proofreading.

- Gemma Derrick, for her insights on impact.

- Amy Eckert, for her positivity and encouragement.

- Charlotte Fleming (ireadyourwriting.co.uk), for her helpful and amusing editorial comments.

- Nathan Hall, for his academic humour and for the crowdfunding boost.

- Jason McDermott for the amazing RedPen BlackPen comics illustrating the book.

- Grainne Kirwan, for giving me my first opportunity to give a silly lecture at a real university.

- Ivan Oransky, for the support and the great work at Retraction Watch.

- Raul Pacheco-Vega, for his invaluable contribution to the academic community, and for taking the time to send me some kind words at the perfect moment.

- Jens Persson, Pontus Böckman, and all at the Skåne branch of the Swedish Skeptics Society, for pledging to have me to speak at their monthly meeting.

- Julia Pierce and the team at Scrivener (literatureandlatte.com), for their incredibly generous contribution to the crowdfunding effort, and for developing the excellent software that I used to organise my scatterbrain into a book.

- Kat Peake, for her keen eye, writing companionship, and dislike of overenthusiastic italicisation.

- Graham Steel, for the encouragement and good humour.

Convention obliges me to self-effacingly declare that any errors or omissions are my own. They are not. If you spot any, please alert me immediately so I can work out who is to blame.

People are important, but so are places (especially as lately I seem to spend more time in cafes and libraries than in the office). I therefore wish to acknowledge: the Anticafé in Paris, where I started the weekly Shut Up and Write session, and where the blog and book started to take shape; the coffee shops of Brno, Czech Republic, where I did a substantial chunk of the initial drafting in one intensive week alongside my good friend Johannes; and the coffeeshops of Amsterdam, where Emily kindly paid for me to accompany her on a work trip, during which I hashed out the final draft.

Despite considerable evidence to the contrary, the following people thought I wasn't completely off my trolley and generously contributed a significant amount of money to the crowdfunding effort that made this book possible: Mark Archibald, Chris Ashford, Dawn Bazely, Ondrej Cernotik, Rebecca Dunn, Simon Haslam, Elyse Ireland, Andy Franklyn-Miller, David Graham, Paul Miller, Neville Morley, Jussi Paasio, Debi Roberts, Deborah Roberts, and Bradley Turner.

I am incredibly grateful to you all. May your papers be published, may your students read the syllabus, and may your editors leave your Oxford commas unmolested.

TABLE OF CONTENTS

X.   CONCLUSION

TABLE OF FIGURES

TABLE OF TABLES

TABLE OF INSTITUTIONALISED SEXISM ♀

AN ACADEMIC BLESSING

May the tenure track rise up to meet you.
May your deadlines always be extended.
May your teaching load be lightened,
the reviewers fall soft upon your papers,
and until we meet again
(Wednesday at the interminable weekly faculty meeting),
may the Dean hold you
in the palm of His hand.

WHAT'S ALL THIS NONSENSE THEN?

Contemporary academia could be seen as a hothouse
for functional stupidity.

Alverson and Spicer, 2012[1]

Academia. Stuffy middle-aged men sporting elbow patches. Greying mad scientists, slightly muddle-headed and socially incompetent. Grand buildings with dusty halls and libraries, sinking beneath the weight of arcane books.[*] Elderly professors skateboarding around campus, cats publishing physics papers in French, and conference presentations consisting entirely of the word 'chicken' repeated over and over.

If academia is a world apart, the unusual aspects of it that I am about to show you take place in an altogether different dimension. I drifted into this strange place by accident. The first day I sat down in my PhD office, ready for three years[†] of hard research and writing (not to mention social

---

[*] Despite urban legends to this effect circulating amongst students since at least the late 1970s, there is no evidence that this has ever really happened.

[†] Five years and counting.

isolation and financial instability), I hadn't a clue what I was supposed to be doing. I wasted much of the first week watching cat videos on the internet and playing inane games on my phone.*

I started researching in earnest around week three. Ten or so pages into the search results for 'marine energy'† I came across a completely irrelevant (for the purposes of my dissertation) paper entitled 'Energy Saving Through Trail Following in a Marine Snail'.[2] Naturally, I was intrigued. I proceeded to read the article in its entirety, learning that the marine intertidal snail (*Littorina littorea*) can achieve an energy saving of approximately 75% by following the trail previously laid by a fellow snail. I also learned, albeit indirectly, that academics are researching the most random of subjects.

I created a folder entitled 'Obscure' alongside all the serious stuff and stashed away the snail paper. I frequently added further fodder to the folder.‡ Not only was it a fun way to procrastinate, but occasionally dipping into the entertaining tit-bits I had collected kept me grounded, reminding me of the (in)significance of my actual research.

It wasn't until much later that this minor folly turned into something approaching an obsession. One evening in Paris, in conversation with my good friend Bart, I remarked that I would eventually write a book about the bizarre side of academia. He told me that nobody would read it, so we made a wager. The fact that you are reading this attests to the failure of his hypothesis (thank you).

Before that fateful conversation, social media had always brought out my inner Luddite, but I swallowed my pride and created a blog and accompanying Twitter account. *Academia Obscura* was born (and a significant portion of my free time was lost forever).

Academics were evidently in need of comic relief because the project proved popular in a way that I hadn't expected. This probably shouldn't

---

* I wish that were a joke.

† My PhD research looks at the legal and regulatory aspects of wave and tidal energy technologies, sometimes collectively referred to as 'marine energy'.

‡ Always avoid alliteration, alternatives are available.

have come as a surprise. Academic work can at times be unexciting and isolating – we need a collective outlet for our frustrations, and humour has often played this role. As James McConnell (founder of the *Worm Runner's Digest*, one of the first academic parody publications) put it:[3]

> *Humour in a scientist, a sort of controlled lunacy, serves as a safety valve that ensures that he remain intellectually open.* ♀

The relationship between humour and academia is nonetheless fraught. There are, broadly, two camps: those who think that jokes and humour have no place in science and academic inquiry; and those who think that we should all just lighten up a bit.[4] I am predictably (and staunchly) in the latter category. One academic, of the former disposition, responded to one of my crowdfunding emails: 'Dear Glen, Strangely enough, I'm not keen to fund a book that rubbishes my job in such a one-sided way.'*

It is true that misguided attempts at humour occasionally backfire. The French scientists deliberately naming various genetics processes so as to spell out *'Ta mère en string panthère'*† come off as humourless at best (and as middle-aged white guys making cringeworthy and immature sexist jokes at worst).♀[5] This book is about the stuff that's not just puerile, but actually amusing.

Academic humour assumes many forms: hoaxes, spoofs, satirical journals, silly science experiments, etc.‡ I've also found, and will share with you, sham 'scientific' journals that are so outlandish they seem

---

* I felt bad, so I replied to apologise for the uninvited intrusion into his inbox and politely explain that I did not want to rubbish academia. He wrote back: 'I apologise for condemning without reading it first. Always a mistake! All the best'. (But he still didn't pledge for the book.)

† Loosely translated as 'F**k your mother in a leopard-skin G-string'.

‡ 'Etc' is the abbreviation academics use when they can't think of further examples but want to give the impression that they have plenty left up their sleeve.

satirical, inadvertently amusing errors and faux pas, plain bad manners, and excessive eccentricity from those who should know better.

The Ig Nobel Prizes, the awards that celebrate creative research that 'first makes you laugh, then makes you think', are undoubtedly one of the most recognisable outlets for academic humour. The Igs, organised by Marc Abrahams under the umbrella of the *Annals of Improbable Research*, are almost as popular as the real Nobels – around 9,000 nominations are sent in each year. The *Annals* itself follows in a long line of parody publications, dating back to the late 1950s when a number of such periodicals first began poking fun at the peculiarities of the academy (including *The Journal of Irreproducible Results* and the *Worm Runner's Digest*).

There are also more muted attempts to inject humour into the academic enterprise, like the jokes and jibes that academics slip into their otherwise serious peer-reviewed papers when they think nobody's looking. Authors citing porn stars and football teams as sources of inspiration, listing Muammar Gaddafi as their co-author, or including this illustration of a rat in pants:[*]

Fig. 1. The underpant worn by the rat.

Figure 1: The underpant worn by the rat

---

[*]   See page 200 for more details.

Most of the examples in this book are unique and absurd one-offs that are unlikely to be repeated. But I have been driven to wonder how many isolated instances one needs to observe before concluding that a significant portion of the academic community is, in fact, slightly unhinged.

The internet has allowed these oddities to garner a greater share of eyeballs than previously possible, precipitating a bold new era of academic humour. Jokes once buried deep in papers only to be uncovered by a handful of curious researchers are now liable to be spotted and spread rapidly, while school scandals and dodgy dealings are exposed in a heartbeat. At the same time, ticked-off professors and PhD students can now find a global community with whom they can vent their frustrations and share stories. Social media accounts like Shit Academics Say reach an audience numbering in the hundreds of thousands, spreading their unique brands of scholarly sarcasm and snark far and wide.

Like all good academic works, I shall start out with the caveat that the scope of this book is limited. The flow of academic antics is constant, and the seam of strange runs surprisingly deep. It is simply not possible to cover every quirky bit of nonsense. I am constrained by space and time (space-time?) to present only the finest selection of academic obscurities.

I probably should be writing something 'useful' or finishing my PhD, but I have had such fun with *Academia Obscura* that I feel it would be a shame not to share it.

My ulterior motive is that I will never again struggle to respond to the question, 'What do you actually do?', or even worse, 'Have you nearly finished your thesis?' Instead, I will just present the questioner with a copy of this book and hope that they are sufficiently baffled to never bother me again.

If you are yourself an academic, I hope that you will do the same and that this book inspires you to take academia a little less seriously. If you are not an academic, I don't pretend that this book will even begin to explain what academics do, but I hope it will make the mass of impenetrable papers and lofty conferences seem more accessible, bring a smile to your face, and inspire you to take us a little less seriously too.

PUBLISH OR PERISH

'Publish or perish' is at once the academic's motto, curse, and raison d'être. The well-worn adage is etched into the brains of grad students and professors alike. It is hard to pin down the exact provenance of the phrase, though it seems that it has been in use since at least the early 1940s when Logan Wilson wrote:[1]

> *The prevailing pragmatism forced upon the academic group*
> *is that one must write something and get it into print.*
> *Situational imperatives dictate a 'publish or perish' credo*
> *within the ranks.*

Universities and funders are now placing increasing emphasis on alternative means for disseminating research, and have expanded the focus on publishing papers[*] to a range of other 'P's' – presentations, project proposals, postdocs, PhD supervision. Nonetheless, publications remain the hard currency of academia.

---

[*]   Publishing in the academic context generally means writing a paper for a peer-reviewed academic journal: you write the manuscript and send it to a journal; they get a couple of your peers to read it and give you feedback before publishing it. (See pages 37 and 51 on scholarly publishing and peer review respectively.)

So ingrained is publish or perish in the academic psyche, we often continue to publish even after perishing. Alfred Werner, the first inorganic chemist to win the Nobel Prize, published a paper in 2011, notwithstanding his death in 1919. His fellow Nobel laureate in chemistry, Robert Woodward, was so prolific during his life that the pace of his scientific discoveries outstripped his ability to publish, such that much of his work was published only after his death. One physics paper (mentioned later for its incredibly long list of authors) is notable for the fact that twenty-one of the co-authors were no longer alive at the time of publication.

Should you have the misfortune to spend any length of time reading academic papers, you will notice common elements: title, abstract, acknowledgements, methods, discussion, conclusions, footnotes, etc. Spend as long as I have looking at academic papers, and you will notice that each element is an opportunity for academic humour: a snide comment, an Easter egg,* or a massive mistake that is only uncovered years after publication.†

WHAT'S IN A NAME?

Because the title of a paper is the first thing the reader sees, it's important

---

* An 'Easter egg' is a hidden message, inside joke, or feature (usually in video games and other interactive media). Though not the first example, the term was coined in 1979 to describe a hidden message in the Atari video game *Adventure*. Programmer Warren Robinett knew that his employer didn't include programmers' names in game credits (because they were worried that competitors would poach their employees), so he secretly inserted a credit that would only be displayed if the player hovered over a single grey pixel in a particular part of the game. The message was found only after he had left the company. The director of software development, Steve Wright, realised that reprogramming the game would be costly, so he reframed the incident and encouraged future games to include such messages as 'Easter eggs' for players to find. The insertion of such Easter eggs have been common ever since (e.g. Go to Google and search 'do a barrel role' or 'askew').

† Or worse, a mistake with the potential to sink your career that people notice instantly.

that it gives them a clear sense of what to expect. However, academics tend to do the opposite, using unfamiliar words and expressions, (mixed) metaphors, or questions. It often feels like authors have carefully chosen their titles to be as obfuscating as possible.

The titles that irk me most are those that awkwardly use tired clichés in an attempt to enliven the subject matter and entice the reader.* I have seen countless papers claiming that one thing is dead, so long live another thing, while topics that have been described as a 'perfect storm' range from 'alcohol and caffeine' to 'sleep in adolescents'.[2] As a researcher on ocean issues, I've seen a lot of 'rising tides' and 'shifting sands'.† I'll concede that 'Leading a Sea Change in Naval Ship Design' and 'Missing the Boat on Invasive Species' are apt uses of maritime metaphors, but 'A Rising Tide Meets a Perfect Storm: New Accountabilities in Teaching and Teacher Education in Ireland' is a bridge too far.[3]

One of the earliest studies of such titles was written by Philip Atkin for the 2002 Christmas issue of the *British Medical Journal*.[4] The issue is dedicated to spoofs and parodies, which explains Atkin's apparent enthusiasm for clichés: 'Papers with catchy titles work best. Titles need to contain phrases that are in popular use and suggest innovation and exploration.' The paper analyses the use of 'paradigm shift' and 'pushing the envelope', both popular clichés at the time. He found 201 papers during

---

* The word cliché is onomatopoeiac from French: it was the sound a movable type printing plate made when it was in use. Given that letters were set individually, it made good sense to cast frequently used words and phrases as a single piece of metal. Over time, cliché came to mean such a ready-made phrase, and eventually took on the meaning it has in English today.

† While we tend to use 'rising tide' to refer to a growing number or trend, it first caught on after John F. Kennedy used the phrase 'a rising tide lifts all boats' to express the idea that improvements in the general economy will benefit everyone, and therefore economic policy should focus on macroeconomic development (though really he was trying to justify a pork barrel project he was inaugurating – the Greers Ferry Dam). Though commonly attributed to JFK, the phrase was originally the slogan for a regional chamber of commerce, the New England Council, and was repurposed by Kennedy's speechwriter Ted Sorensen.

the period 1976–2001 that contain the former, and 37 the latter. 'Paradigm shift' was initially unpopular, but that shifted in the mid-1980s. A period of exponential growth followed, but the phrase suffered a steep decline as the noughties approached. Likewise, academics were pushing few envelopes early on, but then in the 1990s we started to give them a real beating.

With presumably sarcastic exuberance, Atkin urges academics to use new and exciting words and phrases in paper titles: 'We must not confine our meditations but should begin to think outside of the box.'

Ten years later, Neville Goodman revisited Atkin's work and found that 'paradigm shift' had rebounded, while mercifully few envelopes were being pushed.[5] Atkin's nod to thinking outside the box was prescient: the phrase first appeared in 1995 and 124 papers used it in the period 2006–10.

Table 1: Frequency of clichés used in medical article titles (1971–2010)[*]

Cliché	Year of first usage	#
State of the art	1959	3518
Gold standard	1979	915
Paradigm shift	1980	722
Cutting edge	1970	411
Outside the box	1995	200
Wind of change	1960	184
Coalface/Goalposts/playing field	1990	164
Quantum leap	1972	48
Rubber hits the road	1985	23

[*] Adapted from Goodman's paper. Goodman based his analysis on searches in PubMed, a database focused on medical fields. Global numbers would likely be much higher.

To be or not to be?

Clichés are only the tip of the iceberg. Goodman conducted another study of titles, 'From Shakespeare to *Star Trek* and beyond: A Medline Search for Literary and Other Allusions in Biomedical Titles'.[6] He found over 1,400 Shakespearean allusions, a full third of which are to 'What's in a name',* and another third to *Hamlet*.

'Much ado about nothing' appears 171 times, the first in 1967 as 'Much ado about the null hypothesis',[7] while the 'be' in 'to be or not to be' has been substituted for a range of other things. 'To Clone or Not to Clone' appeared in 1997, one year after the successful cloning of Dolly the sheep.†[8] 'To Test or Not To Test' is used over 3,500 times, including some gems like 'To test or NOD-2 test: what are the questions?'[9] Peak Shakespeare was reached in 'Breast Cancer Screening: All's Well that Ends Well, or Much Ado About Nothing?'[10]

Beside the Bard, Goodman found 244 allusions to Hans Christian Andersen's *The Emperor's New Clothes*.‡ According to academia, the emperor has a motley wardrobe containing everything from isodose curves to 'the lateral ligaments of the rectum'.[11] One paper references both Andersen and Shakespeare ('Mentorship – Is It a Case of the Emperor's New Clothes or a Rose by Any Other Name?'),[12] while 'Evidence-Based Practice: Sea Change or the Emperor's New Clothes?' simultaneously pushes my ocean cliché button and ticks the Andersen box.[13]

Goodman argues that such paper titles are a learned behaviour and that we are likely to see new allusions creep into titles over time. Sadly, he seems to be correct. Authors are already playing around with 'Winter

---

* 'What's in a name? That which we call a rose by any other name would smell as sweet' from *Romeo and Juliet*.

† Dolly got her name from the fact that the somatic cell from which she was cloned was derived from a mammary gland cell and that the scientists 'couldn't think of a more impressive pair of glands than Dolly Parton's'.

‡ Even after discounting papers about emperors or emperor penguins.

is coming' (a quote from *Game of Thrones*), though even here there is the occasional chuckle-worthy effort – e.g. 'Winter is Coming: Hibernation Reverses the Outcome of Sperm Competition in a Fly'.[14]

The Good, the Bad, and the Ugly

Swiss science journalist Reto Schneider has been documenting the use of films as paper titles.[15] The 1968 spaghetti western *The Good, the Bad and the Ugly* is the clear frontrunner, with around 2,700 publications substituting the 'ugly' with everything from 'the whole grain' to the 'Cell Type-Specific Roles of Hypoxia Inducible Factor-1 in Neurons and Astrocytes'.[16] Remixes of *Sex, Lies and Videotape* are also frequent, though considerably less salacious in the academic incarnation 'Sex, Lies and Insurance Coverage' (which discusses legal liability for the negligent transmission of sexually transmitted diseases).[17]

The majority of film allusions are contrived. 'Everything You Always Wanted to Know about Amorphophallus, but were Afraid to Stick your Nose Into!'[18] will make sense to a botanist,* but I don't see why you'd be afraid to ask questions regarding protein kinases.[19] Likewise the exclamatory tone of the title 'Honey, I Shrunk the Article! A Critical Assessment of the Commission's Notice on Article 81 (3) of the EC Treaty' no doubt belies the arcane contents within.

Of Mice and Men

Nobody has yet taken on the considerable task of documenting references to classic novels in paper titles, though there are likely thousands. Biochemist Eva Ansen weaved 41 paper titles alluding to Steinbeck's *Of Mice and Men* into a poem, producing some riveting rhyming couplets:[20]

---

* The sexual organ visible on the plant bears more than a passing resemblance to that of the human male.

Of mice and men: the evolving phenotype of aromatase deficiency.
Of mice and men: an introduction to mouseology or, anal eroticism
 and Disney.[*21]

Certain classics lend themselves to lazy exploitation: *A Tale of Two Cities* can become a tale of two pretty-much-anythings, from organisations to auto plants;[22] a *Catch-22* might present itself to anything from special education reform to 'amphibian conservation and wetland management in the upper Midwest'.[23]

Plug the title of any classic into your academic search engine of choice for literally hours [minutes] of fun.

Like a Rolling Stone

As part of a long-running bet, five Swedish scientists have been sneaking Bob Dylan lyrics into paper titles. This is how a paper on intestinal gases acquired the title 'Nitric oxide and inflammation: The answer is blowing in the wind'.[24] Elsewhere, the Rolling Stones have been immortalised ('"I can't get no satisfaction": The impact of personality and emotion on postpurchase processes'),[25] as have ABBA ('Money, money, money: not so funny in the research world')[26] and Nirvana ('Smells Like Clean Spirit').[27] A paper providing a history of rock in the 1990s has the apposite subtitle, 'A stairway to heaven or a highway to hell?'[28] Though Goodman found no 'Fat-Bottomed Girls' at the time of his 2005 study, just a year later a paper on the mating habits of spiders was published entitled 'Female morphology, web design, and the potential for multiple mating in *Nephila clavipes*: do fat-bottomed girls make the spider world go round?'[29]

Shit Happens

Vaguely intellectual Shakespeare allusions aside, occasionally authors simply have an urge to indulge their immature inclinations. I imagine

---

* This is a real paper. I know because I read it. I'd have finished writing this book months earlier had I not been constantly tempted to read all of the strange studies my research turned up.

that the respective authors of 'An In-Depth Analysis of a Piece of Shit' and 'Shit Happens (to be Useful)!' giggling to themselves as they pressed the submit button.[30] Likewise, the authors of a study proving that a 'hyperbolic 3-manifold containing large embedded balls has large Heegaard genus' say at the end of the paper's introduction: 'A proper subset of the authors wished to subtitle this paper "Big balls imply big genus", which is indeed the best way to memorize the result.'[*]

One View of the Cathedral

Paper titles sometimes make more sense in the context of an ongoing discussion among authors. In 'Write when hot – submit when not' the authors argue that academics would be best advised to submit papers during the winter (as journals tend to receive fewer submissions during this period).[31] The response of James Hartley (author of the seminal *Academic Writing and Publishing*) is entitled 'Write when you can and submit when you are ready!' (which is, in my humble opinion, the better advice).[32]

A shining example of both an ongoing conversation and an overwrought allusion has been with me since my undergraduate years. During a course on Law and Economics, we studied a 1972 paper entitled 'Property Rules, Liability Rules, and Inalienability: One View of the Cathedral'.[33] The subtitle references a series of paintings by Monet of the same cathedral (in Rouen, France) in a variety of lighting and weather conditions, the implication being that the paper offered only one of several perspectives.

The paper has garnered around 2,700 citations, and many other authors have built on the cathedral metaphor.[34] I don't doubt that Monet would have been capable of painting a 'clear view' or a 'downwind view' of the cathedral. He could possibly have painted a 'better view' (though I wouldn't be the one to critique his artistic abilities), or he might've missed a particularly enticing perspective. In another time he might have taken an 'experimental view' of the cathedral, painted it in a 'different light', or

---

[*]   This pithy summary would make a perfect nanopublication (see page 95).

focused on its shadow. But I am sure that even Monet would have struggled to paint an 'ex ante view', much less a 'feminist critique' of the cathedral.

The lead author of the original paper, Guido Calabresi, praised the 'Simple Virtues of the Cathedral' some 25 years later, but in 'Another View of the Quagmire'[35] Daniel Farber inadvertently summarises the titling saga: 'it is better to get a clear view of the swamp rather than to fool ourselves into believing that there is a cathedral buried somewhere beneath the muck'.

Table 2: Miscellaneous papers with silly titles

Title	Content
'Raeding Wrods With Jumbled Lettres: There Is a Cost'[36]	Tsteed sutdnets on thier raednig seped for txtes wehre wrods had jumbled lettres. Unsurprisingly, it is harder to read jumbled words.
'Not guppies, nor goldfish, but tumble dryers, Noriega, Jesse Jackson, panties, car crashes, bird books, and Stevie Wonder'[37]	Explores the so-called 'Guppy effect', i.e. that some conjunctive concepts are typically associated with the conjunction rather than with either of its constituents (e.g. we tend to think of a guppy as more of a pet fish than either a pet or a fish).
'From Urethra With Shove: Bladder Foreign Bodies. A Case Report and Review'[38]	Case report of an 82-year-old man who ended up in hospital after a pencil he was inserting into his urethra broke off inside. Introducing himself to hospital staff, he said he felt 'funny down there'.
'You Probably Think This Paper's About You: Narcissists' Perceptions of Their Personality and Reputation'[39]	Examines whether narcissists are aware that other people perceive them negatively (they are).
'Local Pancake Defeats Axis of Evil'[40]	I have no idea, but I'd watch the movie.

AUTHORS

Some academics are blessed with superb surnames with which to adorn their papers. I am repeatedly confronted by people joking that I am 'Mr (W)Right', and, while I sincerely look forward to appending 'Dr' to my own moniker, I shall forever envy Dr Badger (Dr Boring, less so). There is a plant scientist called Dr Flowers,[41] and two uncanny coincidences come from the world of food science: Ron Buttery has studied the chemical composition of the flavour of popcorn, and Kevin Cheeseman wrote a paper on the fungi used in cheesemaking.[42]

Some amusing author names are entirely accidental. An unfortunate digitisation error caused Antonio Delgado Peris to be rendered as 'A. Delgado Penis' in online databases (*delgado* means 'thin' in Spanish),[43] while the spine of the *Encyclopedia of Animal Behavior*, edited by Michael Breed and Janice Moore, reads:

> *EDITORS*
> *BREED*
> *MOORE*

Academics have also intentionally subverted author lists with surprising regularity. In 1987, physicist William G. Hoover added a fictitious colleague, *Stronzo Bestiale*, to the author list on a paper (Italian for 'total asshole'),[44] while Andre Geim (the only scientist to have won both an Ig Nobel and a real Nobel)[*] listed his hamster, Tisha ('H.A.M.S. ter Tisha'), as his co-author on a paper.[45] When *Physical Review Letters* started allowing authors to transliterate their names into Mandarin, they probably didn't expect that Caltech's Victor Brar would be known as 韦小宝 (Wei Xiaobao)[46] – Wei is the antihero in the Chinese novel *The Deer and the Cauldron*, a prodigal son of a prostitute and a demi-emperor with eight wives.

---

[*] The former for levitating a frog using incredibly strong magnets, the latter for the invention of graphene.

Try as they might, I doubt any academic, human or otherwise, will ever top one Dutch scientist (and winner of the 2011 Name of the Year Award): Taco Monster.[*][47]

CO-AUTHORING: BECAUSE WRITING IS HARD

Choosing your co-authors is not dissimilar to choosing a life partner (except you can always change your partner, but once your name is on a paper, there's no taking it back). Generally, academics team up with colleagues or others from their field, but the literature also evidences some unexpected collaborations.

David Manuwal, an emeritus professor at the University of Washington, managed to get his wife, daughter and son involved in a paper.[48] David's wife had a background in forest ecology, so she sampled plants, his daughter had learned how to identify birds and helped to conduct bird surveys, and his son assisted in marking out the study sites. David's dedicated team carried out their studies in the snowy depths of Washington State in April at temperatures of about −10°C. David claimed it was 'hard work, but enjoyable' (it is not known whether his family share this sentiment).[49]

Four unrelated authors with the surname Goodman collaborated to produce a joke paper entitled 'A Few Goodmen: Surname-Sharing Economist Coauthors'.[50] Similarly, 284 authors sharing the name 'Steve' contributed to a paper entitled, 'The Morphology of Steve'.[51] The paper was a by-product of the National Center for Science Education's 'Project Steve', a comic riposte to creationist groups that had been assembling lists of 'scientists who doubt Darwinism' to cast doubt on the theory of natural selection.[†]

---

* Despite being listed as the author on 13 papers, I couldn't find a university profile for Taco, so I am inclined to think this is a long running and well-executed joke. However, Dutch parents do occasionally call their kids Taco. In 1974, at peak Taco, 58 newborns were given the name.

† A 'spectacularly dumb idea . . . science is not decided by plebiscite'.

The Center assembled a list of scientists called Steve and made T-shirts proclaiming: 'Over 200 scientists named Steve agree: Teach Evolution!' The 284 Steves featured in the paper had all bought the T-shirt, and in doing so had unwittingly given over data regarding their geographic location, sex (the study includes Steve cognates such as 'Stephanie'), and shirt size. The four lead authors (only one of which is called Steve) say: 'We discovered that we had lots of data. No scientist can resist the opportunity to analyze data, regardless of where that data came from or why it was gathered.'*

While 300 authors may seem unmanageable, even for a spoof study, the number of individuals supposedly contributing to academic papers is increasing exponentially. In 1963, Derek de Solla-Price predicted that by 1980 the single-author paper would become extinct. We are now well into the noughties and single-author articles persist, but we have witnessed unfettered growth in author numbers and the emergence of the era of 'hyperauthorship'.[52]

I have personally co-written papers with 15 co-authors, and anywhere between two and ten authors seems to be commonplace. Some papers have taken such collaboration much further, e.g.:

- A paper on fruit fly genomics boasting over 1,000 authors.[†][53]

- A 2016 paper in *Autophagy* with close to 2,500 authors, including 38 Wangs.[54]

- The 2012 paper announcing the observation of the Higgs Boson at CERN with 2,924 authors (the standard practice

---

* They also note: 'the fourth through 443th authors were not consulted concerning the use of their names in this article. They can thank us at their leisure. After all, they are now co-authors with Stephen Hawking and Nobel laureates Steven Weinberg and Stephen Chu.'

† Initially I assumed that the fruit flies themselves made the author list. On further enquiry, I learned that Sarah Elgin, the researcher at Washington University in Saint Louis, Missouri who led the study, decided to credit all those involved. This included over 900 undergraduate students that she enlisted to help with minor tasks. Elgin herself appears last in the author list.

when citing such a paper is to cite the *ATLAS Collaboration* as the author – unlucky for Mr G. Aad of Aix-Marseille Université, who would otherwise have been first in the list).[*]

- A subsequent 2015 paper from CERN involving two of its research teams for the first time resulted in 5,154 authors (the first nine pages contain substantive discussion of the findings; the following 24 are dedicated to listing the authors and their affiliations).[55]

While journals tend not to print such abnormally long author lists in the hard copies, *Physical Review Letters* gave the 5,154 authors of the 2015 CERN paper the pleasure of seeing their names in print. Aside from the serious questions about what 'authorship' even means in such contexts, this is a colossal waste of paper (and/or disk space). Robert Garisto, an editor at the journal, said that the biggest problem with preparing the manuscript for publication was merging the author lists, as each of the teams had their own slightly different styles.[56]

Another challenge is remembering the names of all the contributors. In one *Nature* paper, a research group overlooked no fewer than five authors.[57] They also mispelled some names and mixed up their funding sources. Getting published in *Nature* can be a career-defining moment, so I can imagine the disappointment of the forgotten five upon finding that their efforts were not credited. This error was picked up reasonably quickly, whereas it took two years for anybody to notice a couple of missing co-authors on a paper in *Ecology Letters*.[58] A lead author that overlooks collaborators can perhaps be forgiven, but one has to question the extent of the contribution of a co-author who fails to notice their own absence from an author list.

---

[*] As a Wright, my sympathies lie with Mr V. Zychacek of the Czech Technical University, who is presumably also relegated to the end of most author lists.

The Alphabet Paper

In 1948, Ralph Alpher, then a physics PhD student, and his supervisor George Gamow, wrote a paper entitled 'The Origin of Chemical Elements' (the paper made a weighty contribution to our understanding of the early universe).[*] The paper was due for publication on 1 April, which may have been what spurred Gamow to add the name of his friend, physicist Hans Bethe, to the author list. The late addition meant that the author list read Alpher, Bethe, Gamow, a play on the Greek letters alpha, beta, and gamma.[†]

The paper came to be known as the 'Alphabet paper' and Gamow later explained:[59] 'It seemed unfair to the Greek alphabet to have the article signed by Alpher and Gamow only, and so the name of Dr Hans A. Bethe (in absentia) was inserted in preparing the manuscript for print. Dr Bethe, who received a copy of the manuscript, did not object, and, as a matter of fact, was quite helpful in subsequent discussions.'

Alpher himself was unhappy with the joke, reasoning that the inclusion of another eminent physicist would overshadow his own contribution and that he wouldn't receive due recognition for his discovery.

He was right. There was a flurry of interest in Alpher's findings, and he found himself defending his thesis in a room packed with 300 spectators. Among them were reporters, who latched on to his comment that primordial nucleosynthesis of hydrogen and helium had taken only 300 seconds and ran headlines like 'World Began In 5 Minutes'.[60] Academics showed interest in Alpher's work, he got fan mail, and religious fundamentalists even prayed for his soul.[‡][61]

However, the spotlight soon faded and, as he feared, his role in the

---

[*] The paper tried to show that the Big Bang model of creation could explain the abundances of the light elements in the universe. Though the original theory neglected some key processes in the formation of heavy elements, later developments showed that the basic theory was essentially correct.

[†] Of one R. C. Herman, who contributed to calculations made in the paper, it was said that he 'stubbornly refuses to change his name to Delter'.

[‡] That's when you know you've really made it.

discovery was ultimately overshadowed by his illustrious co-authors, as fellow physicists wrongly assumed they were responsible for the substance of the paper. Even today, Alpher's role is usually overlooked, and he has been dubbed the 'forgotten father of the Big Bang'.[62]

Croquet, anyone?

It doesn't matter whether you have two or two hundred co-authors: as soon as you move beyond one, the question of the order in which the names appear rears its ugly head. I used to assume that common sense would suffice, but, for all their intelligence, eggheads often don't have this in abundance.

Authorship credit tends to be doled out based on the amount of work put in, the contribution made to the final paper, or according to who came up with the core ideas. In one 1989 paper, it is pragmatism and honesty that prevail, as the authors admit that: 'Order of authorship was determined by proximity to tenure decisions.'[63] This is not unheard of: in one survey of 127 papers, four determined author order by proximity to tenure decisions.[64]

Materials scientist (and Twitter funny man) Sylvain Deville has meticulously documented a host of unorthodox methods for determining author order.[65] Randomisation is common, with authors being listed alphabetically, arbitrarily, or, as one paper states, 'in a fairly arbitrary manner'.[66] At least 15 papers state that the order was decided by coin toss. Some of them even specify the type of coin: a two-pence coin in one case, and a weighted coin in another. In one paper, a computer-simulated coin was used, while another specifies that the coin flip took place 'in an expensive restaurant'.[67] Bearing the telltale signs of a sore loser, one paper tells us that author order was determined 'by a flip of what [Dr X] claimed was a fair coin'.[68]

Some authors choose what Deville calls the Galaxy Quest method ('Never give up, never surrender!'), whereby author order is determined by the effort expended on final revisions. (This makes total sense to me as I find this unfortunate necessity to be the most tedious part of the writing process.)

In their paper, Hassell & May state: 'The order of authorship was determined from a twenty-five-game croquet series held at Imperial College Field Station during summer 1973.'[*][69] Not described in the paper are the somewhat underhand methods used to ensure their victory in such tournaments:[70]

> *Croquet was played every lunchtime during May's summer visits on a pitch customised by a large population of rabbits. Visitors were invited to play though inevitably lost due to the huge home-team advantage knowledge of the pitch's precise topography afforded. Visitors also frequently declared themselves disadvantaged by the alleged tactic of being asked complex ecological questions mid-stroke. This was a different game from the traditional English vicarage-lawn contest!*

Some of the especially esoteric methods are difficult to decode:

- Randomly with the S-plus sample function.[71]

- By random fluctuation in the euro/dollar exchange rate.[72]

- Alpha-posed that people compare the sizes of betically.[73]

- By relative exactitude of Bayesian priors.[74]

Others have used less sophisticated methods: a tennis match; rock, paper, scissors; or even 'a scramble competition for peat-flavoured spirit'.[75]

---

* If you are not au fait with croquet, you can learn all about it from Joseph Strutt's 1801 book, titled: *The Sports And Pastimes Of The People Of England From The Earliest Period, Including The Rural And Domestic Recreations, May Games, Mummeries, Pageants, Processions And Pompous Spectacles, Illustrated By Reproductions From Ancient Paintings In Which Are Represented Most Of The Popular Diversions.* (I don't know when we stopped giving books such delightfully excessive titles, but the sooner we get back to that the better.)

ABSTRACTS

Abstracts – the one-paragraph summaries provided at the start of academic papers – are not particularly fertile ground for academic humour. There are, however, some stunning examples of brevity and clarity. The inquisitive title of a 2011 paper 'Can apparent superluminal neutrino speeds be explained as a quantum weak measurement?' is immediately answered by the indifferent abstract: 'Probably not',[76] while the title of the paper 'Guaranteed Margins for LQG Regulators' is contradicted by its abstract: 'There are none.'[77]

The shortest possible abstract was achieved in 1974 by a paper entitled, 'Is the sequence of earthquakes in Southern California, with aftershocks removed, Poissonian?'[78] The abstract simply reads: 'Yes.' Two years later, a second team tried to attain the glory of a one-word abstract with 'Nobody' in response to the title, 'Who Needs More Than Four Quarks?'[79] Unfortunately, it appears that the editors made them add a more conventional abstract just before publication. The second one-word abstract finally came in 1992, in 'Does the One-dimensionalising Model Show Intermittency?'. The abstract reads: 'No.'[80]

Graphic abstracts

A few journals now allow authors to add graphical abstracts to their papers. In a joke that rather missed the mark, a research group mapping the proteomes of various substances posted graphical abstracts that smack of sexism. ♀*

The first that caught the attention of the scientific community was in a paper mapping the proteome of coconut milk, entitled 'Harry Belafonte and the Secret Proteome of Coconut Milk' (Belafonte sang a song in 1957

---

* The proteome is the entire set of proteins expressed by a genome, cell, tissue or organism.

called *Coconut Woman*).[*81] The authors included as their graphical abstract a photo of a topless woman holding a pair of coconuts in front of her breasts.[†] A similar paper on the proteome of honey includes a picture of two women in black dresses holding a bass guitar.[82]

Rajini Rao, a professor at Johns Hopkins University, wrote a polite email to author Pier Righetti (who also happened to be on the journal's editorial board) requesting that the offending images be removed. Righetti responded: 'I wonder if you have been trained in the Vatican. As you claim to be a professor of Physiology, let me alert you that this image is physiology at its best!'[83]

When an author is reluctant to acknowledge wrongdoing, one might expect the editor to take responsibility. Instead, the journal's editor, Juan Calvete, followed up with a textbook non-apology. Concerning himself primarily with his distaste for the unwelcome publicity the scandal had brought him, Calvete was quick to point out that he personally didn't consider the image sexist, and that the authors and editors didn't intend them to be either.

Calvete lamented that the scandal was detracting from his precious lab time, but nonetheless found time to write extensive comments on blog posts covering the incident.[84] In one such comment, he asked whether nude paintings hanging in the Musée d'Orsay are not also sexist. Reading the rest of Calvete's troll-like comments makes it hard to believe that he is the editor of a serious scientific publication and not an angry teenage keyboard warrior.

As this book shows, there is no shortage of subtle and smart ways to inject a bit of humour into an otherwise fun-free zone. This is not one of them.

---

* Noting that Belafonte was 'a great singer and a staunch defender of civil rights and democracy', the authors dedicate their work to him. They also acknowledge that improved understanding of the proteomics of coconut milk would probably not have resulted in any changes to his lyrics.

† I searched for the photo and found that it was lifted (without attribution) from a list of 'The Sexiest US Bartenders'.

FOOTNOTES

> *There are two distinct types of footnotes. There is the*
> *explanatory or if-you-didn't-understand-what-I-said-in-the-*
> *text-this-may-help-you type. And there is the probative or*
> *if-you're-from-Missouri-just-take-a-look-at-all-this type.*
>
> Fred Rodell[85]

We are fast approaching peak footnotes. In his history of this overused and much maligned writing convention, Anthony Grafton laments, 'Most students of footnotes, in recent times, have come to bury, not to praise them …'[86] We can glean three important nuggets of information from this quote: 1. Academia is sufficiently saturated that it is possible to be a student of footnotes; 2. We are fed up with footnotes; and 3. Shakespeare's influence is as strong at the foot of the page as in the titles at the top.

Footnotes are the bane of academic writing. Often they are strewn so liberally across the page that they effectively create a shadow paper, necessitating countless hours of time and effort to format and edit according to whichever style guide the journal happens to demand. For readers, they can be an irritating distraction, making the pages feel longer and pulling tired eyes away from their thread.*

An article in the *Telegraph* crowned Paddy Ashdown the 'King of the Footnote Bores', noting that his 'boring footnotes occasionally refer to other footnotes, which turn out to be even more boring'.†[87] One academic joked, 'I plan someday to write a scholarly article consisting of a single sentence and a twenty-page footnote.'[88] They obviously don't realise that this is already the norm in legal scholarship (especially US law reviews,

---

* Like this.

† For example, a footnote on page 371 in volume 1 of his diaries states: 'For discussion of different kinds of Proportional Representation, see footnotes on p.381.'

where the unwritten rule in that footnotes should take up double the amount of page as the substantive text).[89]

For all their failings, footnotes can be beautiful, as any reader of Terry Pratchett or David Foster Wallace can attest. Occasionally, academic footnotes pass muster too. The first chapter of Bock et al.'s statistics textbook is entitled, 'Stats Starts Here', a footnote to which reads:

> *This chapter might have been called 'Introduction,' but nobody reads the introduction, and we wanted you to read this. We feel safe admitting this here, in the footnote, because nobody reads footnotes either.*

In a mathematics paper, Lara Pudwell recounts an 'elegant proof' to a mathematical problem put forward by one T. J. Kaczynski (i.e. the Unabomber). A footnote to his name reads: 'Better known for other work.'[90]

A PICTURE PAINTS A THOUSAND WORDS

The drab graphs and figures that grace the pages of academic papers rarely add much excitement, though there are some whimsical exceptions, such as this figure from a physics textbook:[91]

Figure 2: Well-prepared cat

Many of the amusing figures in academic papers are disgusting or disturbing, presumably included by the researchers for their shock value rather than for reasons of scientific rigour:

- A paper looking at how long it takes mammals to pee features a close-up of an elephant penis in full flow.[*][92]

- A similar investigation of 'dripping urination by small animals' includes a photo of the lesser dog-faced fruit bat making use of the technique.[93]

- A paper on 'spontaneous ejaculation in a wild Indo-Pacific Bottlenose Dolphin' includes a video still of the crucial moment.[†][94] (The spontaneous ejaculation lasted just under half a second, while an aftershock a few seconds later lasted 0.73s, after which the dolphin 'gently swam away'.)

- 'Float, Explode or Sink: Postmortem Fate of Lung-breathing Marine Vertebrates', an investigation of whale carcass explosions, includes a still from a video of a beached whale bursting.[‡][95] The photo, which features a man running towards the camera with the explosion in the background, is reminiscent of a scene from an action movie.

---

[*] All mammals above three kilograms in weight empty their bladders over a period of 13–21 seconds.

[†] The paper claims to be the 'first report of spontaneous ejaculation by an aquatic mammal'. I would not have doubted the veracity of this claim, but the authors' reassurance made me suspicious. After an extensive search I have been unable to find any other reports of spontaneous ejaculation in marine mammals (there are, however, numerous studies reporting spontaneous ejaculation in rats, cats, mice, hamsters, guinea pigs, mountain sheep, warthogs, spotted hyenas, horses, and chimpanzees. There is also one report of a man that spontaneously ejaculated upon defecation as a side-effect of the antidepressants he had been prescribed).

[‡] The researchers hypothesise that such explosions could explain skeletal disarticulation seen in the fossil record, but conclude that probably isn't the case.

- The infamous paper on homosexual necrophilia in ducks (see page 199) includes an image of the disturbing act.

- 'Fellatio by Fruit Bats Prolongs Copulation Time' discusses the unusual behaviour of female short-nosed fruit bats, *Cynopterus sphinx*, which regularly lick their mate's penis during copulation.* The paper is accompanied by a video of the act in question, complete with cheesy music.†

- A later paper, 'Cunnilingus Apparently Increases Duration of Copulation in the Indian Flying Fox, *Pteropus giganteus*',[96] continues this line of inquiry, including a similarly voyeuristic video.

Occasionally there are figures that appear to have been drawn by people like me, whose artistic inclinations never surpassed shaky stickmen and who struggle to write their own name on the whiteboard. A paper investigating the distribution of hookworm eggs in human faeces is especially notable in this regard for its crude diagram of the stool-collection process.[97]

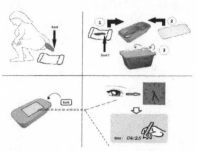

Figure 3: The stool collection process

---

* The researchers found a positive correlation between fellatio duration and copulation duration, with each second of fellatio increasing total sexy time by six seconds.

† The paper also got one of the authors into trouble. He discussed the paper with a female colleague, who later reported him for sexual harassment. He was sanctioned by his university, though an independent investigation found that he was not guilty of sexual harassment. He claimed the sanction cost him tenure and later pursued the university in the High Court. The judge found that the sanctions had been disproportionate.

The title of the paper 'Remains of Holocene Giant Pandas from Jiangdong Mountain (Yunnan, China) and their Relevance to the Evolution of Quaternary Environments in south-western China' scarcely prepares the reader for the storyboard depiction of a poor panda falling off a cliff and slowly rotting into bones.[98]

Figure 4: Possible taphonomic scenario resulting in the accumulation of giant panda bones in the lower chamber

A little more light-hearted, 'Pressures Produced when Penguins Pooh' includes a delightful diagram detailing exactly how far pint-sized chinstrap penguins can shoot their poop.[99]

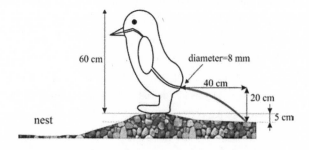

Figure 5: Pressures produced when Penguins pooh

My favourite figure of all time, however, is 'The underpant worn by the rat', so good that it merited inclusion in the introduction. (The author also did the study with dogs, making them wear polyester underpants continuously for 24 months. Sadly, he did not include images or diagrams of the dogs wearing said pants in the paper).*[100]

OOPS

Finding typos in a paper post-publication is dismaying, if inevitable. Even after sinking hours of labour into it there are bound to be some miner errors. This isn't usually fatal and will generally go unnoticed. References to 'screwed data' and a 'screwed distribution' have not stopped a 2004 paper in the *International Journal of Obesity* from garnering over 300 citations.[101] Likewise, a group of Japanese researchers concluded: 'There were no significunt differences in the IAA content of shoots or roots between mycorrhizal and non-mycorrhizal plants.'[102] The paper has racked up 22 citations in spite of the significant slipup.†

---

* This is a shame, because a debate has long raged on the internet as to how dogs would wear trousers, i.e. whether they would be four-legged or two-legged.

† AltMetric (a service that attempts to measure the broader impact of papers) tallies 23 tweets citing the paper – I thought this was pretty decent, until I realised almost all of them are retweets saying, 'Worst. Typo. Ever.'

An unintentionally honest method appears in another paper, where the authors state: 'In this study, we have used (insert statistical method here) to compile unique DNA methylation signatures.'[103]

A couple of cringeworthy blunders have drawn the attention of the academic community in recent years. The Gabor scandal started when an internal author note was accidentally included in the final published version of an ecology paper.[104] The relevant passage reads as follows:

> *Although association preferences documented in our study*
> *theoretically could be a consequence of either mating*
> *or shoaling preferences in the different female groups*
> *investigated (should we cite the crappy Gabor paper here?),*
> *shoaling preferences are unlikely drivers of the documented*
> *patterns ...*

The corresponding author said that the comment was added following peer review during the revision process and unfortunately slipped through the cracks in subsequent rounds of editing. He told Retraction Watch: 'Neither myself nor any of the co-authors have any ill-will towards any other investigators, and I would never condone this sentiment towards another person or their work . . . I apologize for the error.'[105] Caitlin Gabor also got in touch with Retraction Watch and told them that she knows some of the authors, and had previously written a paper with one of them.

A similar mix-up shook the chemistry world in 2014. Due to an error in the editing process, an internal note in the papers supporting information appeared on the journal's website. In the note, the first author appears to have been asked to fake data:[106]

> *Emma, please insert NMR data here! where are they? and for*
> *this compound, just make up an elemental analysis ...*

Elemental analyses are readily fabricated and are easy to slip into a paper if the journal does not ask for a copy of the independent laboratory report.[*][107] In the Emma case, however, the journal ultimately found no evidence of falsified analyses.[108]

Not being a chemist, I am reluctant to pass judgement on those caught up in the scandal. However, I do have considerable sympathy for Emma, especially as substandard practices may not be so unusual anyway (see page 118). One of the founding editors of *PLOS Medicine*, Virginia Barbour, notes that, while the case is unusual in how it came to light, 'questions on data in papers after publication are very common'.[109] I'm not the only one who feels for Emma. After one kind-hearted academic took to their blog to express sympathy and defend Emma, her mother commented on the post:[110]

> *We know that fabricating data would be alien to her.*
> *I cannot believe that her good reputation, built up over*
> *these years can be destroyed in a week. I know nothing of*
> *the academic community, but the hostile and aggressive*
> *comments left on the blog sites are unbelievable. I don't know*
> *if Reto Dorta was careless or has done a very bad thing, but*
> *I do know that Emma is the innocent party in this affair.*

Rest assured that it is not only researchers who make mistakes. The London School of Economics once sent an email to around 200 students to confirm that they had accepted their place at the university, but due to an administrative error the email was addressed to Kung Fu Panda. This error caused some concern in a school where 25% of students are Asian, but apparently the choice of name merely reflected one staff member's fondness for the film. Other names in the test database included Piglet, Paddington, Homer, Bob and Tinkerbell.

---

* This was a central issue in the much publicised 2011 case of Bengü Sezen, a former Columbia University chemistry student who conducted an elaborate fraud to get her PhD.

OBSCURE INTERLUDE:

ACADEMIC WHIMSY

The Bee's Knees: A paper in *Biology Letters* reported that buff-tailed bumblebees choose which flowers to harvest based on the colours of flowers and where they are located relative to each other.[1] But the real revelation is that the paper was written by a class of twenty-five school children from Blackawton Primary School in Devon:

> *We discovered that bumblebees can use a combination of colour and spatial relationships in deciding which colour of flower to forage from. We also discovered that science is cool and fun because you get to do stuff that no one has ever done before.*

If you need a break from a stressful schedule or the abstruse language of academic papers, this one will remind you that at its best, science can be accessible, engaging, and fun for all ages.

Pooh Problems: Another paper, 'Pathology in the Hundred Acre Wood: a Neurodevelopmental Perspective on A. A. Milne'[2] goes in somewhat the opposite direction, taking something joyous and beloved from our childhood and ruining it entirely. The paper takes a look at the dark underside of *Winnie the Pooh* and finds 'a forest where neurodevelopmental and psychosocial problems go unrecognized and untreated'. Piglet suffers from a generalised anxiety disorder, while Tigger has a recurrent pattern of risk-taking behaviours. The prognosis for Pooh is pretty grim: attention deficit hyperactivity disorder (ADHD); impulsivity (evidenced by his misguided plan to wangle honey by disguising himself as a cloud); obsessive-compulsive disorder (OCD); and Tourette's.

The Others: Continuing in the vein of familiar topics viewed through a new lens, 'Body Ritual among the Nacirema' satirises anthropology's tendency to exoticise 'other' cultures by turning the spotlight on the USA.[3] Horace Miner's 1956 paper, published in *American Anthropologist*, provides an introduction to American culture using the vernacular (and occasionally condescending tone) of an anthropologist describing a hitherto uncontacted tribe. Miner focuses on the American obsession with appearance and hygiene, including a 'mouth-rite ritual' that involves 'inserting a small bundle of hog hairs into the mouth, along with certain magical powders, and then moving the bundle in a highly formalized series of gestures.' We are also introduced to the Nacirema's charm-boxes (medicine cabinets), household shrines (bathrooms), medicine men (doctors), and their cultural hero Notgnihsaw known for, amongst other things, 'the chopping down of a cherry tree in which the Spirit of Truth resided'.

Star Man: In 1978, 30 years before winning a Nobel Prize, Paul Krugman wrote a paper entitled 'The Theory of Interstellar Trade',[*][4] ('a serious analysis of a ridiculous subject, which is of course the

---

[*] A footnote to Krugman's name says that the research was supported by a grant from the *Committee to Reelect William Proxmire*, a US Senator that, to put it mildly, was not a huge fan of NASA. He was particularly opposed to space exploration, cutting it from NASA's budget entirely, and effectively ended NASA's nascent 'search for extra-terrestrial intelligence' (SETI) efforts. Proxmire inevitably drew the ire of space advocates and science fiction fans, and Arthur C. Clarke attacked him in the 1960 short story 'Death and the Senator'. Proxmire issued his trademark 'Golden Fleece Award' once a month between 1975 and 1988 to focus media attention on projects he considered self-serving or wasteful. Scientists even began using his name as a verb, meaning to obstruct scientific research for political gain (e.g. 'Our project has been proxmired'). Proxmire was also a fitness buff and wrote a book entitled *You Can Do It!: Senator Proxmire's Exercise, Diet and Relaxation Plan* (1973). The cover is predictably hilarious.

opposite of what is usual in economics' – his words, not mine). Krugman proposes a method for calculating interest on goods that travel at close to the speed of light and proves two 'useless but true theorems' about the dynamics of interest rates in interplanetary markets.

FIVE OUT-OF-THIS-WORLD *STAR WARS* PAPERS

1. It's a Trap: Emperor Palpatine's Poison Pill[5]

Abstract: 'In this paper we study the financial repercussions of the destruction of two fully armed and operational moon-sized battle stations ("Death Stars") in a four-year period and the dissolution of the galactic government in *Star Wars*.'

Highlights: Estimates that the Death Star cost $193 quintillion (including R&D); concludes that the Rebel Alliance would need a bailout of 15–20% of Gross Galactic Product to mitigate the fallout of Death Star destruction.

2. Using *Star Wars*' supporting characters to teach about psychopathology[6]

Abstract: 'The pop culture phenomenon of *Star Wars* has been underutilised as a vehicle to teach about psychiatry . . . The purpose of this article is to illustrate psychopathology and psychiatric themes demonstrated by supporting characters, and ways they can be used to teach concepts in a hypothetical yet memorable way . . . Characters can be used to approach teaching about ADHD, anxiety, kleptomania and paedophilia.'

Highlights: Jar Jar Binks as the 'low-hanging fruit of psychopathology', a uniquely academic (over)analysis of Luke's familial relations.

3. Evolving Ideals of Male Body Image as Seen Through Action Toys[7]

Abstract: 'We hypothesised that the physiques of male action toys . . . would provide some index of evolving American cultural ideals of male body image . . . We obtained examples of the most popular American action toys manufactured over the last 30 years. We then measured the waist, chest, and bicep circumference of each figure and scaled these measurements . . . We found that the figures have grown much more muscular over time...'

Highlights: The accompanying image showing how buff Hans and Luke became between 1978–1998; concludes that they've grown from average blokes to bodybuilders over the last 20 years, with impressive, if unsightly, gains in the shoulders and chest.

4. The Skywalker Twins Drift Apart[8]

Abstract: 'The twin paradox states that twins travelling relativistically appear to age differently to one another due to time dilation. In the 1980 film *Star Wars: Episode V – The Empire Strikes Back*, twins Luke and Leia Skywalker travel very large distances at "lightspeed". This paper uses two scenarios to attempt to explore the theoretical effects of the twin paradox on the two protagonists.'

Highlights: Luke is 638.2 days younger than Leia.

5. That's No Moon

Abstract: 'This article aims to investigate the first "Death Star" from the *Star Wars* film series and how much energy it would require to destroy a planet.'

Highlights: You need 2×10^{27} J to blow up a simplified planet; the Death Star could destroy small- to medium-sized planets, but would not be able to destroy stars.

ACADEMIC PUBLISHING

The academic publishing model is insane. Academics, often funded by the taxpayer, write papers and submit them to journals, which recruit other academics to peer review the work (for free). The publisher lightly edits and formats the paper and posts it online. They then charge the same academics that write and review the papers upwards of 20 quid to read them. Researchers get no royalties or payment, and generally have to sign over copyright as a condition of publishing.

But it wasn't always this way. The first publication resembling the journals we now know and love (to hate) was the *Journal des sçavans*, first published on Monday, 5 January 1665.* Contents included obituaries of famous men, church history, and legal reports. The journal *Philosophical Transactions of the Royal Society* followed a few months later on 6 March 1665.

In those early days of enlightenment, people were growing increasingly curious about the natural world and the laws that governed it. It was fashionable for the aristocracy to be interested in science and, as a result,

---

* The journal ceased publication in 1792, during the French Revolution, and, although it briefly reappeared in 1797 under the updated title *Journal des savants*, it did not re-commence regular publication until 1816. It continues to be a leading academic journal in the humanities.

science was becoming *cool*.* The Royal Society, for example, was formed during the mid-1660s when a group of (yes, old white-haired) men got together to talk about how the world worked. One of them, presumably in a moment of wine-fuelled inspiration said something like, 'Hey, is anyone writing this stuff down?!' and academic publishing was born.

Thus journals began as a way for scientists to share their observations and anecdotes with the world. This was the first time that the weird and wonderful (two-headed calves and the like) were woven into an academic discourse, instead of simply being held up for entertainment.

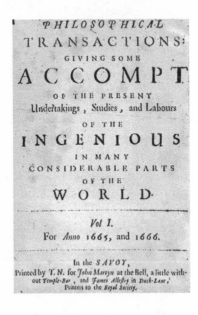

Figure 6: Cover of the first issue of *Philosophical Transactions*

The subsequent proliferation of journals has been unrelenting, particularly in the internet age. There are somewhere in the region of 30,000 journals

---

* Hooray! No more being burned alive at the stake for suggesting that the Earth moves around the sun! Scientists rejoice!

in circulation,[1] and around 50 million articles have now been published.[2] Some journals, like *Nature, Cell,* and *Science* are famous beyond their fields. Most, like the *American Journal of Potato Research*, are not.

While journals were initially driven by an organic curiosity and desire to collaborate (and compete), beginning in the 1960s, commercial publishers began to selectively acquire top-flight journals previously published by not-for-profit academic societies. Because demand for top journals is moderately inelastic, the publishers lost hardly any market share when they jacked up prices. The profitability of these journals drove further consolidation, and now just five companies now account for half of all academic articles published.[*][3]

These companies have eye-watering profit margins (no surprise given that their two primary inputs, the papers and peer review, are provided free of charge) and, while publishers argue that they add value, a 2005 analysis by Deutsche Bank concluded: 'If the process were truly as complex and costly as the publishers claim, 40% margins wouldn't be available.'[4] A 2016 study compared the final published versions of papers to the preprint versions posted online to see if the publication process had changed anything.[†][5] The majority of the papers were exactly the same, so the obvious question is: what are we paying for?

MONEY FOR NOTHING

In what initially appeared to be a brazen example of publishers raking in unearned profits, an accident of digitisation led publishers to charge £20 for 'papers' consisting solely of a single page with the text 'This page

---

*   Reed Elsevier, Springer Science+Business Media, Wiley-Blackwell, Taylor & Francis, and Sage.

†   A preprint is simply a draft of a scientific paper that has not yet been published in a peer-reviewed journal. The timely distribution of a preprint allows the authors to get feedback from their colleagues and peers before formal peer review, acceptance and publication. Preprints are now largely distributed online rather than as paper copies, giving rise to huge preprint databases.

is intentionally left blank'. Once the mistake had been spotted, a flurry of tweets ensued, and two days later I and four other procrastinating academics had written an in-depth analysis of the phenomenon.

The paper 'This Study is Intentionally Left Blank: a Systematic Literature Review of Blank Pages in Academic Publishing' became the third most read paper on *Figshare* in 2014 and was later published in the *Annals of Improbable Research*.

We studied 24 of the 56 Individual Blank Pages (IBPs) found on *ScienceDirect*, finding only one that was truly blank. The rest all contained the stock phrase, furnishing the reader with 31 characters at a cost of approximately $1.33 per character.

While the paper was really just an opportune moment to have a dig at the publishers, blank pages do present an interesting philosophical conundrum: the purpose of the text is to indicate that the page is purposely bereft of content, yet the inclusion of the text means that the page is no longer truly blank.[*]

Graham Steel, one of the co-authors, brought the paper to the attention of Elsevier's head of open access during the 2014 UKSG Annual Conference and Exhibition.[†6] The representative said they had not seen the paper but would take a look. The blank pages soon disappeared, but we uploaded them, making them publicly available to ensure that future research into IBPs can take place unencumbered.[‡7]

---

[*] We posit, inter alia, that intentionally blank pages could be a kōan, i.e. a statement used in Zen practice to provoke the 'great doubt' and test a student's progress.

[†] UKSG originally stood for United Kingdom Serials Group, but as it no longer covers only the UK or serials, the acronym is a touch outdated. The body aims to encourage the exchange of ideas on scholarly communication, and the annual conference is part of this mission, bringing together librarians, publishers, intermediaries, technology vendors, and, occasionally, funny (and slightly angry) Scottish open access advocates.

[‡] This action is, of course, in violation of copyright. Nobody has yet sought their removal, presumably because they would look utterly ridiculous doing so. I personally hoped we would receive a takedown notice and that the Streisand effect would propel our silly study into academic stardom.

[This page is intentionally left
99.855% blank.]

[The page on which this statement has been printed has been intentionally left devoid of substantive content, such that the present statement is the only text printed thereon.]

I have since found another publisher willing to rent you a blank page at $6 for 48 hours, and another charging $40 for access to its 'Instructions for Authors' page. They better be some damn good instructions.

THE REBELLION

Slowly but surely, academics are beginning to challenge the insanity of academic publishing. The open access model, whereby anyone can read the paper without having to surmount a paywall, has rapidly been gaining ground in recent years.

If you'd never heard of open access, you'd be forgiven for thinking that the idea of allowing researchers to access research is so obvious that there shouldn't need to be a movement to support it. Yet academics, afflicted with Stockholm syndrome, have long acquiesced to the status quo. While some publishers are tentatively trying out new models, perhaps aware that popular opinion is turning against them, the majority have understandably been reluctant to engage with anything antithetical to their profitable business model.

Against this backdrop, websites like SciHub are surreptitiously making papers available for free on a massive scale, and researchers are posting their publications online in huge numbers ('Even though technically it's in breach of the copyright transfer agreements that we blithely sign, everyone knows it's right and proper').[8] As academics move to reclaim publishing, publishers are scrambling to claw it back. Elsevier started asking Academia.edu to take down posted publications, and has sued the creator of SciHub, Alexandra Elbakyan, for copyright infringement.

RECOMMENDED JOURNALS

If the end of academic publishing is nigh, you may wish to get your career-defining papers published before the journals go extinct. Here are eight that you might consider:

The American Journal of Potato Research*

In addition to the usual full-length articles, AJPR welcomes 'short communications concisely describing poignant and timely research'. The only poignant thing about the journal is its social media presence: just 80 Twitter followers.†

Rangifer: Research, Management and Husbandry of Reindeer and Other Northern Ungulates

The 'world's only scientific journal dealing exclusively with biology and management of Arctic and northern ungulates, reindeer and caribou in particular' – still going strong after 37 volumes.

Journal of Negative Results in BioMedicine

Called 'the world's most boring journal' by the *Washington Post*,[9] the JNRBM is one of the few journals offering scientists a chance to publish the research that didn't work, combating the ingrained tendency to publish only positive results. As a result, the journal contains lots of false starts and failed hypotheses, such as 'False rumours of disease outbreaks caused by infectious myonecrosis virus in the whiteleg shrimp in Asia' and 'The female menstrual cycle does not influence testosterone concentrations in male partners'.[10]

The Journal of Universal Rejection (JofUR)

The JofUR removes all doubt from the submission process: your paper will be rejected. Sometimes rejection will 'follow as swiftly as a bird dropping from a great height after being struck by a stone', other times it may languish in the editor's inbox, but 'it will come, swift or slow, as surely as death. Rejection.'[11]

JofUR's website suggests some reasons why you might want to submit anyway:

---

* You say potato, I say *Solanum tuberosum*, and that's why academics don't get invited to dinner parties.

† Show them some love, follow @potatoresearch.

- No submission anxiety: you know 100% that it will not be accepted.

- No publication fees.

- One of the most prestigious journals (as measured by acceptance rate).

- Authors retain complete rights over their submitted work.

- A decision is generally reached within hours of submission.

- You can submit whatever you like, however you like ('You name it, we take it, and reject it. Your manuscript may be formatted however you wish. Frankly, we don't care.')

Proceedings of the Natural Institute of Science (PNIS)

In the likely event of rejection by the JofUR, PNIS might take the manuscript ('We'll Publish Anything!' exclaims the website).[12] Claiming to be part serious (I am not sure which part), this satirical journal publishes science funnies in two streams: PNIS-HARD (Honest And Reliable Data) and PNIS-SOFD (Satirical Or Fake Data). Recent publications include a paper investigating whether prayer can help academics attain statistical significance (it can't)[*13] and a paper entitled 'Effects of climate change, agricultural clearing, and the sun becoming a red giant on an old growth oak-hickory forest in southeastern Iowa'.[†14]

---

* 'On one hand, praying before generating a dataset resulted in more significant differences than reciting random text. On the other hand, praying did not perform better than simply doing nothing. Plus, praying had no effect on statistical significance after the data had already been collected (i.e., the Desperation Scenario).'

† 'In the simulation involving a solar progression into red giant stage, oak-hickory forests were reduced to their elemental constituents and redistributed among the cosmos.'

Answers Research Journal (ARJ)

The ARJ is the only journal I know that openly declares that it will only publish articles that accord with a pre-established hypothesis. The journal, whose moniker masks its ulterior motive, publishes research that: 'Demonstrates the validity of the young-earth model, the global Flood . . . and other evidences that are consistent with the biblical account of origins.' Highlights include a series of articles attempting to estimate the number of various species types aboard Noah's Ark,[*] and extensive guidance on how to reference religious texts properly.[†]

Nursing Science Quarterly

Rosemarie Parse established *Nursing Science Quarterly* twenty-five years ago and remains the editor today. This is not especially unusual. However, Parse herself also appears to be the main topic of the journal, as the majority of published papers cover her own ideas and theories. Parse also founded an eponymous international society and yearly conference, and you can even buy a Parse pin badge. Not many journals boast their own complementary line of jewellery.[‡]

DODGY OPEN ACCESS

Journals publishing open access articles sometimes charge authors a fee to publish, partly in a bid to offset the costs of running a well-oiled journal

---

[*] The papers consist of seemingly scientific language, followed by a load of nonsense based on a comically literal interpretation of the Bible. E.g. On the genus *Acrochordus*: 'because of its fully aquatic existence and capability of osmoregulating in hypotonic and hypertonic aquatic environments, it is potentially capable of surviving Flood conditions and are not included on the Ark'.

[†] E.g. 'Lowercase for divine dwelling places, including heaven, hell, and paradise.'

[‡] Send a stamped-addressed envelope and a blank cheque today to receive an exclusive Academia Obscura tie clip.

Dear Editor,

It is not clear why a cover letter is required except to fulfil the silly British preoccupation with letterhead and other emblems of status.

   Please accept my correspondence.
Sincerely,[*]

---

[*]   I spotted this superbly honest cover letter on Twitter (author unknown).

machine,* and partly (read: largely) to maintain those all-important profits.† The fact that there is money to be made is finally drawing traditional publishers toward open access, but it has also been exploited by unscrupulous actors, turning the model into a potential source of hoaxes and hijinks.

Internet scams used to be something of a blunt instrument: wealthy widows with tax avoidance schemes or wealthy Nigerian princes seeking to surreptitiously shift some cash under the radar. Then came 'phishing' – using social engineering techniques to con people into voluntarily handing over valuable information. When the scammers hit academia, they started to get smart(ish), producing journals and organising conferences to exploit academics eager to add the next line to their CV.

Such journals are generally of extremely low quality, publishing papers with little or no editing or review, deceiving authors about the fees involved, and falsely claiming that high-profile scientists are on the board of editors. They regularly send emails to researchers to solicit manuscripts, often offering generous discounts on the processing or publishing fees and promising a tantalisingly rapid turnaround (without peer review and proofreading, they get the articles out instantaneously).

Junk journals are not a huge problem in and of themselves because the vast majority of experienced academics see them for what they are and refrain from submitting their work or sending money.‡[15] Sadly, the small number of academics sending their work to such journals tend to be young and inexperienced researchers from developing countries.[16]

---

* Just kidding, most journals are still using clunky outdated systems with incredibly inefficient workflows.

† Around two thirds of 'pure' open access journals listed by the Directory of Open Access Journals don't charge a publication fee, but so-called 'hybrid' offerings from traditional publishers (i.e. subscription journals that contain some open access articles) generally involve higher fees.

‡ This is worth noting because traditional publishers have used the issue of junk journals as a PR tool to argue that they are the only ones capable of providing reliable open access publishing, which is patently not the case.

Aware of the increasing number of invitations arriving in his inbox, Jeffrey Beall, an academic librarian and a researcher at the University of Colorado in Denver, started scrolling through the websites of these unknown journals. He quickly realised that many of them, despite sporting grandiose names, were not as scientific as they sounded. Beall started a list of so-called 'predatory' journals in 2010 with 20 entries; the list now runs to 4,000 entries.

Some of the publishers on Beall's list, including the Canadian Center of Science and Education and OMICS, have threatened to sue him for defamation and libel. The threat from the latter was about as exaggerated as the claimed quality of the scientific products being churned out: OMICS said it would seek $1 billion in damages and that Beall could be imprisoned for up to three years under India's Information Technology Act. In a lengthy letter, OMICS argues that Beall's list is 'the mindless rattle of a incoherent person' that 'smacks of literal unprofessionalism and arrogance', and accuses him of racial discrimination. For their part, OMICS recently had many of its journals delisted from a leading publication database, while the US Federal Trade Commission is charging them with deceiving academics and hiding publication fees.[17]

The integrated journal of what now?!

Dodgy journals are simple to spot thanks to their spammy emails. The *Integrated Journal of British* is one such rag. The email sent to advertise the launch of the journal enthusiastically begins: '!! Greeting IMPACT FACTOR: 3.3275'.*[†] There is nothing British about the journal, which

---

* If they were going to be so ridiculous, they could've at least listed their impact factor as pi.

† Though claimed to be a 'verified' impact factor, a quick skim through the list of journals that *Universal Impact Factor* has supposedly accredited reveals that this company is also a thinly veiled sham operation. Rated journals are based in 'Bulagria' and 'Corea', while clicking on a journal title for more information will fill your screen with popup ads for penis enlargement pills and other typical internet junk.

is based in India, or its content. The journal's logo is a wolf surrounded by stars, apparently lifted from the website of a small Wisconsin home improvement company.

Still, I don't think *The Integrated Journal of British* is the worst journal of all time. I would bestow that dubious honour upon the *American Based Research Journal* (ABRJ). Its website declares that it is an 'Open-Access–Monthly–Online–Double Blind Peer Reviewed Journal'. Despite its name, the website lists a UK contact address.*

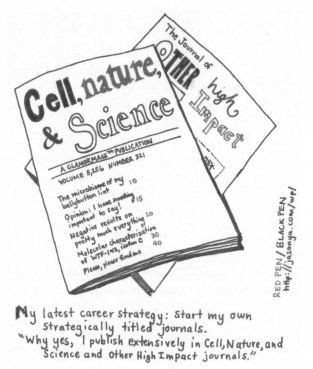

Figure 7: Strategically titled journals

---

*   A house on a nondescript suburban lane on the outskirts of Manchester, just down the road from the Manchester Ukrainian Cultural Centre and the Museum of Transport.

The stated scope of the journal is bewildering, with subjects covered ranging from 'Fundamentals of Income Tax' to 'Fashion Trends'. They chose a stylised DNA double helix as the logo to reflect this dizzying scope. Online publication costs $150, and the journal regularly spams scholars to solicit submissions.

Displaying blatant disregard for both the proofreading and mail merge functions, one of their spam solicitation messages starts: 'Dear Dear Author, We are really impressed after reading your research work: "Research Article".' The email starts bad, gets worse, and is then signed off by the editor 'Dr Merry Jeans'. No matter how many times I've read it, I still chuckle at Dr Merry Jeans. The editorial board of ABRJ features a cast of such comic characters, including 'Dr Belly Joseph', 'Dr Jazzy Rolph', and 'Prof. William' (no surname), none of whom really exist.

Curious to know who was behind this operation, I did some digging. I found that ABRJ's web address is registered to someone based in Lahore, Pakistan, who was previously a student of the Virtual University of Pakistan. His personal blog consists of just one telling post, in which he brags that he has been suspended from university for posting completed university assignments online.

PEER REVIEW

Sticks and stones may break my bones, but it's the withering peer review comments that do the long-term psychological damage.

Peer review is not the prettiest of processes. Regardless of your discipline or the journal in which you publish, one of the reviewers will invariably: 1. Ask you to write a completely different paper (i.e. the paper they would have written); 2. Demand that you repeat or expand expensive and time-consuming experiments; or 3. Reject your paper out of hand, often with demoralising and petty comments.

While peer review is supposed to provide quality control, plenty of

Your manuscript as submitted

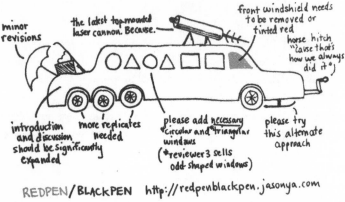

... and after peer review and revision

REDPEN/BLACKPEN http://redpenblackpen.jasonya.com

Figure 8: Your manuscript on peer review

journals are publishing utter rubbish, and there are countless occasions where journals have rejected important results (going back at least as far as the 1796 rejection by *Philosophical Transactions* of Edward Jenner's report of the first vaccination against smallpox).

Academics generally approach peer review as an unfortunate ordeal to be overcome on the road to publication, rather than as the scholarly meeting of minds we nostalgically tell ourselves it might once have been. Rebecca Schuman writes:[18]

Think of your meanest high school mean girl at her most gleefully, underminingly vicious. Now give her a doctorate in your discipline, and a modicum of power over your future. That's peer review.

My personal experience has fortunately been less harrowing. For me, peer review is an underwhelming experience: tiresome and tedious at its worst; mildly helpful at its best. Yet every academic has a sob story or two, and while the vast majority of peer reviews move smoothly, it is inevitably that minute fraction of cruel comments that plagues us.

The baptism of fire I received upon my first paper submission is one such experience. I have long since deleted the rejection email, which had weighed heavy like a horcrux on my inbox, but I recall that it was an outright rejection, followed by a list of reasons why the paper I was trying to write was ludicrously ill-conceived (followed by an even longer list of reasons why I hadn't succeeded in any case).

The appropriately anonymous blog *Shit My Reviewers Say* collects the worst of the worst, while the *Journal of Environmental Microbiology* periodically publishes colourful comments submitted by its reviewers.

Some reviewers are simply hard to please:

- 'The whole paper reminds me of a paper of a couple of years ago, which I didn't like.'
- 'Can you explain this part a bit further, but without going into detail.'
- 'Something is missing.'
- 'Didn't like this one.'
- 'Is there a chance you could send me any good papers, at least once in a while?'

The worst are downright brutal in their rejections:

- 'This paper is desperate. Please reject it completely and then block the author's email ID so they can't use the online system in the future.'

- 'I am afraid this manuscript may contribute not so much towards the field's advancement as much as toward its eventual demise.'

- 'It is early in the year, but difficult to imagine any paper overtaking this one for lack of imagination, logic, or data – it is beyond redemption.'

- 'The work that this group does is a disgrace to science.'

- 'Presumptuous, ignorant and downright dangerous.'

- 'The writing is often arrestingly pedestrian.'

- 'Reject – More holes than my grandad's string vest!'

Occasionally, in their rush to criticise others, reviewers get themselves tongue-tied:

- 'The article could benefit from a good linguistic editing in order for it to be better sound and flowing.'

- 'I was not sure exactly which problem the author is trying to solve and vice versa it was not clear to me what problem the solution is intended to solve or explorer.'

- 'If the paper is accepted, I strongly recommend an English prof-reading.'

Finally, some reviewers return comments so cryptic they seem designed to make the author question their own sanity:

- 'I would refrain from using enumerations in your paper and instead encourage you to think about the deep masculinism that comes with.'

- '650 should be lowercase.'

- 'This needs some rephrasing – it's loaded with the assumption that there is a real world.'

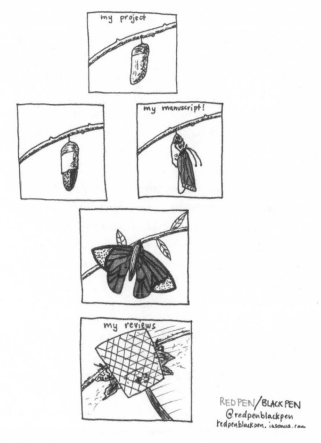

Figure 9: My reviews

A quick whip-round on Twitter turned up plagiarising reviewers accusing the authors of plagiarism (pot, kettle, you know the story), a reviewer that shouted 'THIS DOESN'T EVEN MAKE SENSE', and a reviewer suggesting that a paper written by a native English speaker was obviously not written by a native English speaker and should be proofread by somebody with a proper command of the English language.[19]

There is an occasional side of sexism served up with a rejection ('This paper reads like a woman's diary, not like a scientific piece of work').♀

Fiona Ingleby, an evolutionary geneticist at the University of Sussex, experienced this first-hand when a peer reviewer suggested that she enlist male co-authors to 'serve as a possible check against...her own ideologically biased assumptions.'[*][20] The journal said it would scratch the anonymous academic from their list of potential reviewers.

Frances Healey, Associate Director of Patient Safety at the NHS Commissioning Board Authority, received the following comment from a reviewer:

> *When my son was five we discussed what type of dinosaur we should keep in the garden as a pet. Some might scare the dog, others would eat Mum's flowers. In the end we decided not to have a dinosaur at all. Which more or less sums up this paper. You have put in a lot of effort answering a question that should never have been asked, but you do arrive at a sensible conclusion.*

Responding to such unhelpful peer-review comments is in itself an art form. Frances and her co-authors were both gracious and humorous in their response, which ends: 'We hope our reviewer's son is growing up with his dad's sense of humour, and a real rather than imaginary pet.'

Roy Baumeister of Florida State University composed the following template cover letter for those struggling to be so gracious:

Dear Sir, Madame, or Other:

Enclosed is our latest version of MS# XX-XXX-XX-, that is, the re-re-re-revised revision of our paper. Choke on it. We have again rewritten the entire manuscript from start to finish. We even changed the goddamn running head! Hopefully we have suffered

---

*   Her paper investigated gender differences in the transition from PhD to postdoc, leading the reviewer to comment that: 'It might well be that on average men publish in better journals...perhaps simply because men, perhaps, on average work more hours per week than women, due to marginally better health and stamina.'

enough by now to satisfy even you and your bloodthirsty reviewers.

I shall skip the usual point-by-point description of every single change we made in response to the critiques. After all, it's fairly clear that your reviewers are less interested in details of scientific procedure than in working out their personality problems and sexual frustrations by seeking some kind of demented glee in the sadistic and arbitrary exercise of tyrannical power over hapless authors like ourselves who happen to fall into their clutches. We do understand that, in view of the misanthropic psychopaths you have on your editorial board, you need to keep sending them papers, for if they weren't reviewing manuscripts they'd probably be out mugging old ladies or clubbing baby seals to death. Still, from this batch of reviewers, C was clearly the most hostile, and we request that you not ask him or her to review this revision.

Some of the reviewers' comments we couldn't do anything about. For example, if (as reviewer C suggested) several of my recent ancestors were drawn from other species, it is too late to change that. Other suggestions were implemented, however, and the paper has improved and benefited. Thus, you suggested that we shorten the manuscript by 5 pages, and we were able to accomplish this very effectively by altering the margins and printing the paper in a different font with a smaller typeface. We agree with you that the paper is much better this way.

One perplexing problem was dealing with suggestions #13–28 by Reviewer B. As you may recall (that is, if you even bother reading the reviews before doing your decision letter), that reviewer listed 16 works that he/she felt we should cite in this paper. These were on a variety of different topics, none of which had any relevance to our work that we could see. Indeed, one was an essay on the Spanish–American War from a high school literary magazine. The only common thread was that all 16 were by the same author, presumably someone whom Reviewer B greatly admires and feels should be more widely cited. To handle this, we have modified the Introduction and added, after the review of relevant literature, a subsection entitled 'Review of Irrelevant Literature' that discusses these articles and also duly addresses some of the more asinine suggestions in the other reviews.

We hope that you will be pleased with this revision and will finally recognize how urgently deserving of publication this work is. If not, then you are an unscrupulous, depraved monster with no shred of human decency. You ought to be in a cage. May whatever heritage you come from be the butt of the next round of ethnic jokes. If you do accept it, however, we wish to thank you for your patience and wisdom throughout this process and to express our appreciation of your scholarly insights. To repay you, we would be happy to review some manuscripts for you; please send us the next manuscript that any of these reviewers submits to your journal.

Assuming you accept this paper, we would also like to add a footnote acknowledging your help with this manuscript and to point out that we liked the paper much better the way we originally wrote it, but you held the editorial shotgun to our heads and forced us to chop, reshuffle, restate, hedge, expand, shorten, and in general convert a meaty paper into stir-fried vegetables. We couldn't or wouldn't have done it without your input.

Sincerely,

(your name here)

Another option is to reject the rejection. Two researchers from the University of New South Wales in Sydney provide a template for such a move in the 2015 Christmas issue of the BMJ. Their letter begins:[21]

> *Thank you for your rejection of the above manuscript.*
> *Unfortunately we are not able to accept it at this time.*
> *As you are probably aware we receive many rejections*
> *each year and are simply not able to accept them all. In*
> *fact, with increasing pressure on citation rates and fiercely*

*competitive funding structures we typically accept fewer
than 30% of the rejections we receive. Please don't take this
as a reflection of your work. The standard of some of the
rejections we receive is very high.*

Einstein once rejected a rejection, withdrawing his paper and taking it elsewhere. In 1936 he submitted the paper 'Do Gravitational Waves Exist?',[*22] written with his first American assistant, Nathan Rosen, to *Physical Review*. The editor, John Tate, was unsure of Einstein's conclusions, and sent it to an expert for review. Einstein had not been accustomed to peer review, and was taken aback by the ten-page report picking apart his paper. He wrote back to Tate:[23]

*We (Mr. Rosen and I) had sent you our manuscript for
publication and had not authorized you to show it to
specialists before it is printed. I see no reason to address the
– in any case erroneous – comments of your anonymous
expert. On the basis of this incident I prefer to publish
the paper elsewhere.*

Sometimes, no matter how you respond, there is nothing you can do to change your fate:

Editor comments: *Please respond to Reviewer 2's
comments, who suggested Rejection of the paper.*

Reviewer 2 comments: *None.*

---

* Gravitational waves, the ripples in the fabric of space-time caused by massive dense bodies (like black holes and neutron stars) orbiting each other, were predicted by Einstein in 1916, based on his theory of general relativity. In 2016, the Advanced Laser Interferometer Gravitational Wave Observatory (LIGO) announced the first clear detection of gravitational waves.

The Dawn of Peer Review

By RedPen BlackPen

Editor summary: Ugck-ptha, et al. report the development of 'fire', a hot, dangerous, yellow effect that is caused by repeatedly knocking two stones together. They claim that the collision of the stones causes a small sky-anger that is used to seed grass and small sticks with the fire. This then grows quickly and requires larger sticks to maintain. The fire can be maintained in this state indefinitely, provided that there are fresh sticks. They state that this will revolutionise the consumption of food, defences against dangerous animals, and even provide light to our caves.

Reviewer 1: Urgh! Fire good. Make good meat.

Reviewer 2: Fire ouch. Pretty. Nice fire. Good fire.

Reviewer 3: An interesting finding to be sure. However, I am highly sceptical of the novelty of this 'discovery' as Grok, et al. reported the finding that two stones knocked together could produce sky-anger five summers ago. (I note that this seminal work was not mentioned by Ugck-ptha, et al. in their presentation.) This seems, at best, to be a modest advancement on his previous work. Also, sky-anger occurs naturally during great storm times – why would we need to create it ourselves?

I feel that fire would not be of significant interest to our tribe. Possibly this finding would be more suitable if presented to the smaller Krogth clan across the long river?

Additional concerns are listed here.

1 The results should be repeated using alternate methods of creating sky-anger besides stones. Possibly animal skulls, goat wool or sweet berries would work better?

2 The dangers with the unregulated expansion of fire are particularly disturbing and do not seem to be considered by Ugck-ptha, et al. in the slightest. It appears that this study has had no ethical review by tribe elders.

3 The colour of this fire is jarring. Perhaps trying something that is more soothing, such as blue or green, would improve the utility of this fire?

4 The significance of this finding seems marginal. Though it does indeed yield blackened meat that is hot to the touch, no one eats this kind of meat.

5 There were also numerous errors in the presentation. Ugck-ptha, et al. repeatedly referred to sky-anger as 'fiery sky light', the colour of the stones used was not described at all, 'ugg-umph' was used more than twenty times during the presentation, and 'clovey grass' was never clearly defined.

INTERVIEW

THE SEMI-PROFESSIONAL RANTER

Jon Tennant is a palaeontologist. He rants about things in pubs and thinks this is what science is.

How the hell do you find time to do all this ranting and write a PhD about dinosaurs?

Have you ever tried not having a life? It works wonders for your career. Which is what I'd tell you if I had any semblance of a career. Also, it's crocodiles, not dinosaurs.

But I like dinosaurs. If you were a dinosaur, which would you be and why?

Fukuiraptor. Obvious reasons.

Do you have a mortifying peer review nightmare story?

One time I got Adam Sandler as a referee. He just told me to watch all his movies, made a joke about my mum, and then rejected my paper as it didn't reference *Big Daddy*.

Describe the traditional model of academic publishing in 140 characters.

Shit. That's less than 140 characters, isn't it? Still space? Something something corporate greed.

What is the future of academic publishing?

One that acknowledges that the internet is a thing.

And peer review?

Democratic. Without reviewer 2.

What is your preferred post-coital cheese?

Now now, briehave.

Favourite Twitter hashtag?

#ElsevierValentines

Any bad advice for young academics?

Do everything senior people tell you to do. Being at university is all about conforming to the status quo.

You wrote a cool book – wanna plug it?

It's called *Excavate Dinosaurs*. It has DIY dinosaurs that you pop out and build. I'm happy to plug it because it's awesome, and I don't get royalties, because publishers.

David Tennant: any relation?

According to the restraining order, no.

RETRACTIONS

Even a paper that has passed the rigorous review process may later turn out to be fundamentally flawed. In such cases, a paper can be formally retracted from the literature.[*][24] One of the first English language retractions was self-submitted by Benjamin Wilson to the *Philosophical Transactions of the Royal Society* on 24 June 1756.[†] It reads:[25]

> Gentlemen,
>
> I think it necessary to retract an opinion concerning the explication of the Leyden experiment, which I troubled this Society with in the year 1746, and afterwards published more at large in a Treatise upon Electricity, in the year 1750; as I have lately made some farther discoveries relative to that experiment, and the minus electricity of Mr Franklin, which shew I was then mistaken in my notions about it…
>
> I shall be very glad to have this acknowledgement made public, and to answer that end the effectually, I wish that it may have a place in the Transactions of the Royal Society.

---

* Minor faults may not necessitate a full retraction and can instead be corrected, though the stories behind small corrections are generally not as interesting. One recent correction in *Nature* nonetheless caught my eye. It reads: 'The figure given for the planting of super soya bean in the News Feature "Frugal farming" should have been 67,000 hectares, not 1 million. In addition, the feature failed to make it clear that Jonathan Lynch was joking when he suggested that students should "drop acid".'

† Pinning down the first ever retraction is a difficult task, not least because early uses of the word 'retraction' tended to denote corrections to a paper rather than a full retraction. Science historian Alex Csiszar from Harvard found such an instance dating from 1684, while 5th-century theologian Saint Augustine wrote an entire book of *Retractationes* ('revisions') toward the end of his life to correct everything 'which most justly displease me in my books'. Retractions only started to approach their current format post-WWII.

Wilson had been locked in a public debate with Mr Franklin* on the question of whether lightning conductors should be round or pointed at the top, and had previously arranged an audacious demonstration before King George III at the Pantheon on Oxford Street in London to prove his point.†[26]

Retractions are an important part of the scientific process, yet they generally receive scant coverage. There has, however, been increasing interest in improving documentation of retractions in recent years. Leading the charge is Retraction Watch, once called the 'Garbage Collectors of Science' by a Swiss radio station.[27] Retraction Watch looks out for retraction notices, follows up on tips regarding faulty science, and aims to improve the overall transparency of the scientific publishing process.

Co-founders Ivan Oransky and Adam Marcus say that 'retractions are born of many mothers' and, while outright fraud is quite rare, such cases are especially damaging to both science and to the career of the perpetrator.[28] Anaesthesiologist Scott Reuben spent six months in prison for faking data and was ordered to pay back $360,000 in restitution for misusing grant money.[29] Dong-Pyou Han, a former researcher at Iowa State University, received a 57-month prison sentence and an order to repay $7m in grants after he spiked samples of rabbit blood with antibodies to make a potential vaccine against HIV appear more effective than it truly was.[30]

Caught on camera

Retraction notices posted by journals are typically terse affairs. For example, a notice retracting a 1994 paper from *Nature* simply read: 'We

---

* I.e. Benjamin Franklin, one of the founding fathers of the United States and renowned polymath, author, and scientist.

† His colleagues were not impressed, saying that his 'perverse conduct ... produced such shameful discord and dissensions in the Royal Society, as continued for many years after, to the great detriment of science'. The Pantheon was demolished in 1938 to make way for a new branch of Marks and Spencer (which is still there).

wish to retract this Article owing to an inability to reproduce the results.' Yet the real story is closer to spy vs. spy than science.[31]

Karel Bezouska was one of the foremost biochemists in the Czech Republic, until an ethics committee at Charles University in Prague found that he had probably committed repeated acts of scientific misconduct. In one absurd instance, Bezouska realised that his results could not be replicated, so he broke into a lab where another team was attempting to replicate his results and adulterated the samples in an attempt to change the outcome of the experiments. A student working in the lab tested the samples and found that they'd been handled without authorisation. The lab installed CCTV cameras and caught Bezouska breaking into the room and surreptitiously rummaging around in their fridge.

Fake it until you make it

Faked peer review is one of the more egregious violations of academic integrity leading to retractions. In August 2012, Korean researcher Hyung-In Moon had several papers retracted because he himself had peer-reviewed them.[32] Moon suggested preferred reviewers during the submission process who were either himself or bogus colleagues. In some cases, he simply invented names, but on other occasions he used the names of real researchers (so that a web search would verify their legitimacy) and created email accounts that could be used to provide the peer-review comments. To make the reviews appear more realistic, he submitted favourable comments, but provided some critical feedback or suggestions on how the paper might be improved.

Similarly, in August 2014 SAGE Publishers retracted 60 articles from the *Journal of Vibration and Control* after a 14-month investigation revealed a similar scam.[33] The scandal centred on Peter Chen, formerly of the National Pingtung University of Education in Taiwan, who had created various aliases to enable himself to peer-review and cite his own papers. The publisher admitted that it could not definitively determine the number of individuals involved as their attempts to contact 130 suspicious email addresses resulted in precisely zero responses. The publisher and editor of the journal confronted Chen with the allegations in late 2013. When they were unsatisfied with Chen's explanation, they alerted his University. Chen resigned in February 2014, and in May the editor retired and resigned from the journal. The fallout didn't stop there. Taiwan's then education minister, Chiang Wei-ling, had supervised the thesis of Chen's twin brother and appeared on several of the retracted papers. Ultimately Wei-ling also resigned over the scandal.[34]

Plagiarism

Good old-fashioned plagiarism is no doubt common, but one paper in particular could easily have been dismissed as an April Fool's joke. The *Indian Journal of Dermatology* retracted a paper on plagiarism . . . for plagiarism.[35] The paper included definitions and strategies to detect and prevent plagiarism, but was itself found to have been copied from a master's dissertation. The author of the retracted paper, Thorakkal Shamim, had been part of a panel of experts on plagiarism consulted by a student a few years earlier. Shamim had copies of the responses to a questionnaire the experts had answered and decided to publish the results, spelling mistakes and all, simply adding an introduction and a conclusion. To make matters worse, Shamim had previously taken a hard line on plagiarism, writing an article suggesting that plagiarising authors should be blacklisted and banned for submitting an article for at least five years, and that the head of the author's department and institution must to be notified.[36]

In a similar incident, the author of an article on reincarnation sought to reincarnate the Wikipedia page on reincarnation, copying and pasting considerable chunks of text directly into the manuscript.[37] The retraction notice states that the paper was being pulled because of 'duplicity of text'.[38]

Calling bullshit

A Washington State University investigation found that a researcher studying how to turn cow manure into natural gas fabricated data in a journal article (and also failed to declare a commercial conflict of interest).[39] Rather than admit to the falsification, the researcher told the investigation that he had lost the data. He claimed that a wind storm dumped his notebook into a manure pit during a visit to a dairy farm, and that photocopied pages of the notebook were lost at his sister's house. He neglected to provide an explanation for the loss of all the data files stored on his office computer.

Obese

Peculiar circumstances precipitated the retraction of a paper on obesity treatment from *Biochemical and Biophysical Research Communications*. The authors were all affiliated with the University of Thessaly, a real university in Greece. The authors were however not real: no trace of them can be found online, and the correspondence address is not an official institutional one. The names of the second, third and fourth authors appear to have been sloppily copied and pasted from a real paper,* with the other two being copied from a different paper. Bruce Spiegelman, a cell biologist at Harvard, said that he had presented similar findings at various research meetings and was preparing to submit them for publication. The particular proteins being studied had not previously been the subject of any paper looking at their role in obesity, so Spiegelman was suspicious.

The real intrigue here is why anyone would want to pull such a move. Spiegelman is, in addition to his academic posting, a co-founder of a company developing therapeutics for metabolic disorders, and he reckons that premature publication of his results was a malicious act intended to complicate future patent applications relating to the results.[40] Luckily, Spiegelman had already applied for the patents.

Hearts and minds

Michael LaCour had struck academic gold. His study, entitled 'When contact changes minds: An experiment on transmission of support for

---

*   The names were pasted with the superscript letters denoting author affiliation – i.e. Kapelouzou[c] in the real paper becomes Kapelouzouc in the fraud.

gay equality'[41] challenged the conventional wisdom that attempts to win hearts and minds only entrenches existing views. The paper was a hit on social media, and *This American Life* dedicated a whole podcast to it.[42] I listened with anticipation as the findings were described: a single instance of personal contact with someone affected by the ban on gay marriage could change a person's opinion on the issue. The result seemed too good to be true. It was: LaCour had faked the data.

The unravelling began when Joshua Kalla and David Broockman from the University of California, Berkeley pored over the numbers. Noticing some inconsistencies, they published a damning report describing the multiple reasons they suspected something shady.[43] They realised that the baseline 'feeling thermometer', which was supposed to be calibrated to local samples, was instead identical to a freely available national dataset. In addition, the changes in participants' feeling thermometer scores were perfectly normally distributed – i.e. not a single participant changed their mind in a way that meaningfully deviated from the distribution – a highly unlikely result in the real world.

The researchers reached out to a senior co-author of the paper, Donald Green, to alert him to their discovery. Green agreed that unless LaCour had a good explanation, a retraction was in order. LaCour provided no such explanation. At first, he claimed that he'd simply lost the data. Later, he would claim that he had destroyed the data to comply with privacy and confidentiality protocols.

Green recalls: 'I sent off my retraction, and I went to sleep and I woke up in the morning at 5:30 and there was a lot of email.'[44]

An editorial in the *Wall Street Journal* snidely suggested that the LaCour paper was so popular because it 'flattered the ideological sensibilities of liberals'.[45] As a sensitive liberal snowflake myself, I was certainly happy to hear of the findings, and equally disappointed to learn that people are just as set in their ways as we always knew them to be.

But there is a heartening twist. The two whistle-blowers were themselves in the middle of conducting a similar study, with opinions on transgender people as the subject. Their study found that the canvassing strategy really can change people's minds.[46]

Treefinder

While many are working hard to reduce prejudice, one academic's attempt to further entrench outmoded attitudes led to a 2015 retraction from *BioMed Central*. The journal retracted a highly cited paper describing the software Treefinder (software that creates trees showing potential evolutionary relationships between species) because the lead author and software developer changed the licence terms to make it unavailable in certain countries.[47] Firstly, in February 2015, creator Gangolf Jobb prohibited US users from using the software, citing the country's imperialism. Then in October 2015, he prohibited its use in countries he viewed as too immigrant-friendly, bringing the paper into conflict with the journal's policy that all software discussed in papers be freely available.

Jobb told Retraction Watch that the software is still available to any scientist willing to travel to non-banned countries:

> *Every scientist can use Treefinder, as long as he or she does it in one of the allowed countries and is personally present there. However, having to travel to a neighbouring country is inconvenient, I admit. I don't care.*

His co-authors, who had no say in the decision, readily supported the retraction (though I imagine that losing a paper cited over 700 times must have hurt a bit). Sandra Baldauf, a biologist at Uppsala University in Sweden, was one scientist that was happy to go back to the drawing board: 'I would stop using [Treefinder] just on general principle, even if we had to resort to using pencil and paper.'[48]

Con Man

Diedrek Stapel, a social psychologist from the Netherlands, was something of a star in his homeland. Stapel wrote many well-regarded studies on human attitudes and behaviour, and his results, like those of LaCour, often told us what we wanted to hear (or at least expected to hear) about human nature. Stapel also precipitated his own dramatic downfall by perpetrating a bold academic fraud over the course of a decade, fabricating results and ultimately notching up over 50 retractions.

One of Stapel's much-publicised studies, appearing in *Science*, purported to show that a dirty environment brought out people's latent racist tendencies. Stapel supposedly conducted a study at Utrecht train station that showed that white people tended to sit further away from a black person on a bench when the surrounding area was strewn with litter compared to when it was tidy. Years later, in the midst of the self-initiated unspooling of his career, Stapel visited the train station and realised that there was no location there that matched the fictional one he had meticulously described in the paper.

Stapel has never denied that his deceit was driven by ambition, a common thread among high-flying fraudsters. However, he was also obsessed with order and had long been driven to frustration by what he saw as the imperfect nature of experimental data. Instead of crunching the cumbersome numbers of the real world, Stapel concocted results that were pleasing to the eye. 'It was a quest for aesthetics, for beauty – instead of the truth,' he said in a tell-all interview with the *New York Times*.[49]

Another of Stapel's creations highlights his questionable quest for order. He designed a study to test the hypothesis that people presented with a bowl of M&Ms will eat more if they are primed with the idea of capitalism. Subjects would answer a questionnaire: half would do it sitting in front of an M&M-filled mug emblazoned with the word 'kapitalisme' and the other half would have a mug adorned with jumbled letters. Stapel had a student load the mugs, M&Ms, and questionnaires into his car, saying that he'd conduct the study at a local high school. Instead, he drove home, binned the majority of the questionnaires, and set about simulating the experiment. Eating what he believed to be a reasonable quantity of M&Ms, he filled out the questionnaire and built a dataset around that estimate.[*][50]

The long-running and wide-ranging nature of Stapel's fraud provided the perfect opportunity for a couple of language experts to investigate the linguistic fingerprints of fraud. They analysed patterns in 24 of Stapel's fraudulent papers (170,008 words) and compared them with 25 of his genuine publications (189,705 words). They found that the writing style matched known patterns of deception in language, including, for example, the use of fewer adjectives in fraudulent papers. The fraudulent papers also contained a greater number of words pertaining to methods, investigation, and certainty.[†] This is the painful irony of Stapel's search for perfection: he unwittingly wrote the hallmarks of deception into his otherwise perfect papers.

Foiled

Finally, here is a retraction that was quite close to home. Colleagues at my research institute had recently published a paper about ocean warming and acidification in *Science*[51] when I learned of a conference paper pulled from 'Heat Transfer 2014'.[52] The climate-sceptic author claimed to have single-handedly debunked ocean warming with a home-made experiment using tin foil and cling film.

---

[*] Across the Atlantic, Google's HR team ran an in-house study nicknamed 'Project M&M', wherein they strategically shifted the complimentary candy to opaque containers and instead emphasised the placement of healthy snacks in glass jars. In the New York office, during a period of seven weeks, the 2,000 staff consumed 3.1 million fewer calories from M&Ms.

[†] Stapel also included fewer co-authors when reporting fake data, though other elements of the papers (such as the number of references and experiments) did not vary.

THE GARBAGE COLLECTOR OF SCIENCE

Ivan Oransky is a co-founder of Retraction Watch. He is the vice president and global editorial director of MedPage Today, *Distinguished Writer in Residence at New York University's Carter Journalism Institute, and vice president of the Association of Health Care Journalists.*

How did you first become aware of the world of retractions?

I was deputy editor at the *Scientist* magazine for six years (2002–08). Retractions were rare, but when they happened there was often an interesting story behind them.

How many retractions are there?

Around 500–600 per year, 5,000–6,000 in total, although there were close to 700 in 2015. The rate has gone up dramatically in the last 15 years.

How did Retraction Watch start?

I got to know Adam Marcus, a medical journalist who had broken a few big retraction stories, in particular that of anaesthesiologist Scott Reuben. We'd share details about different cases we saw, about the stories, the ethics and the fallout. I said, 'Let's start a blog,' to which Adam replied, 'Sure, whatever that means.'

What's next?

Our next big project is creating a database of retractions. A lot of people are amazed that one doesn't already exist. You could of course cobble together your own, but you wouldn't have consistency or, importantly, the real reasons for the retractions.

For example, it used to be thought that fewer than half of retractions were due to fraud or misconduct, but we now know that's not the case because estimates were relying on retraction notices. A 2012 paper used RW and other sources to estimate that two thirds of retractions are down to misconduct, which has changed our understanding.[53] The database will allow new work like that to take place and let us analyse patterns.

How are retraction notices misleading?

Retraction notices are often simply unreliable. They vary greatly from journal to journal. Sometimes they say literally nothing, other times they obfuscate the true reasons for the retraction. Overall they don't give a clear picture, so when people look at retraction notices and try to understand the phenomenon, they are likely to be misled.

What's the best retraction notice you've seen?

There are plenty of amusing instances where journals dance around the truth – we have even published a couple of lists of 'plagiarism euphemisms'. Our favourite was a clear case of plagiarism where the journal ventured that this was 'an approach to writing'. Adam commented that this was an approach to writing in the same way that showing up to a bank with a gun is an approach to banking.

Some journals go to great lengths to avoid using the 'p' word. One said that several passages from another paper 'could be viewed as a form of plagiarism', another noted that a paper had an 'originality issue'.

Why so coy?

Journals tell us that lawyers play an outsized role in all of this – they sometimes take an aggressive stance and journals back down because they don't want to deal with excessive legal costs. Accused scientists have been suing institutions, journals, and even commenters on PubPeer (a website that allows users to discuss and review scientific research). We are keeping an eye on it.

Bloody lawyers. What's the future of retractions?

We'd be happy if we didn't have retractions at all, they are the nuclear option. Instead we need a solid correction mechanism and to stop thinking of papers as immutable. Science is an iterative and incremental process and papers should reflect that. There are a lot of initiatives being developed to take this forward, like PubPeer and CrossMark. Nonetheless, I don't think retractions are going anywhere in the near future.

Do you have a personal favourite?

There's always some interesting news or a baffling story that makes it fun for us. I have a favourite category – fake peer reviews. One researcher has notched up 28 retractions because he did almost all of his own peer reviews. His system was ultimately foiled because all the reviews came back in under 24 hours. The editor became suspicious because he did not believe that real reviewers would have turned the papers around so quickly! Those papers probably should have been published anyway, but I guess getting your reviews back in 24 hours with guaranteed acceptance is a pretty good insurance policy.

Is the pressure to publish leading to increased misconduct, or are we just getting better at spotting bad behaviour?

The rise in retractions is dramatic – the rate increased tenfold between 2001–10. However, we must have a sense of perspective. There are millions of published papers, so hundreds of retractions is still not that many. I think the rise is mostly down to the fact that we are getting better at finding misconduct. We now have plagiarism detection software rooting out the most flagrant cases; there are many more readers of papers because everything is online; new tools and communities are poring over papers to find inconsistencies and problems, so it is no surprise that the rate is going up.

I do however think that the pressures on researchers and the incentive structures must contribute in some way: you have to publish papers to get tenure, grants, promotion – i.e. everything you need to have a successful career in science. Everyone does what they think they need to do. For some that means working incredibly hard, a few cut corners, while a tiny minority simply start making things up.

You got a lot of attention during the LaCour scandal. How have such high-profile cases affected Retraction Watch?

We broke the LaCour story, and it had a dramatic impact on us. I got a tip via Twitter – it was very early in the morning and I happened to be awake. Once I confirmed that a senior author was requesting a retraction we broke the story. It crashed our server – I had to pay $300 to upgrade that day to cope with the traffic. The interviews were constant. I did one from South Korea at 1 a.m., then *NPR*, then somewhere else. The *New York Times* profiled us and published our op-ed on the case. It was an incredible boost for us.

Do authors often self-retract?

Self-submitted retractions are not even a large minority yet, but we do have a category on the site called 'Doing the Right Thing'. We try to highlight and praise authors that do self-correct – about a hundred posts so far. Retracting still has a stigma, and no one likes to see their work go to waste, but we think it is better to hear the story from the authors themselves.

Most retractions?

We have a leader board, there are currently around 30 people on it. They shift around as new information comes in. Yoshitaka Fujii is currently number one with 183 papers and shows no signs of budging any time soon. Fujii is a good example of the new tools and communities that are sniffing out bad science – he was caught out by peers who meticulously ran the numbers on his papers.

Funniest retractions?

The paper on plagiarism guidelines that had been plagiarised. Or the two cases where hidden cameras were used to catch researchers tampering with experiments – twice in 2 million papers really does make these one-in-a-million! Best intentions and safeguards will never stop those that are determined to cheat.

On your merchandise page there is a Retraction Watch clock. Can I buy a Retraction Watch watch?
You aren't the first to suggest that ...

Damn, I thought I was being funny. If people want to support Retraction Watch how can they do that?
We appreciate any and all support – reading our site, commenting on posts, sending us tips, telling your colleagues about us. If people are able to make a financial contribution, we are a registered non-profit and they can do so via our site. Thanks for helping us spread the word!

THE HOAXES WITH THE MOSTEST

In 1996, Alan Sokal, a physics professor at New York University, became infamous as the instigator of the best-known hoax in academic publishing history. At the height of postmodernism's popularity, Sokal submitted a paper to the journal *Social Text* entitled 'Transgressing the Boundaries: Towards a Transformative Hermeneutics of Quantum Gravity'.[54] The paper proposed that quantum gravity is a social and linguistic construct, and ostensibly demonstrated how 'postmodern science provides a powerful refutation of the authoritarianism and elitism inherent in traditional science'.[55]

Sokal did not write the paper as a genuine work of critical theory, but as, in his own words, 'A pastiche of left-wing cant, fawning references, grandiose quotations and outright nonsense.'* Sokal wanted to test whether a leading journal of cultural studies would publish an article 'liberally salted with nonsense if it sounded good and it flattered the editors' ideological preconceptions.' The answer was a resounding yes.

The journal did not have a peer-review process at the time, so the paper wasn't reviewed by an external expert, much less a physicist. Sokal revealed his hoax on publication day, igniting a debate about the scholarly merit of humanistic commentary on the physical sciences, as well as on academic ethics (i.e. whether Sokal was wrong to deceive, and conversely whether the journal erred in its lack of academic oversight).

The *Social Text* editors said they thought Sokal was honestly seeking 'some kind of affirmation from postmodern philosophy for developments in his field' and that the paper was a 'change of heart, or a folding of his intellectual resolve'.[56] None of the editors suspected that the piece was a parody, and even once they learned it was a hoax, they argued that it was still of interest as a 'symptomatic document' (i.e. as an example of how awkwardly a natural scientist might approach postmodern epistemology). Sokal was probably further amused that the glaring absurdity was not

---

* Where I'm from, we call this 'bollocks'.

patently obvious. Indeed, in just the second paragraph he claims that physical reality is just a social and linguistic construct. 'Fair enough,' he says. 'Anyone who believes that the laws of physics are mere social conventions is invited to try transgressing those conventions from the windows of my apartment. I live on the twenty-first floor.'[57]

The editors of *Social Text* won the Ig Nobel Prize for literature that year, for 'Publishing research that they could not understand, that the author said was meaningless, and which claimed that reality does not exist.'[58]

The best hoaxes have a serious point to make. Three enterprising MIT graduate students, wanted to expose the daylight robbery that is shoddy academic conferences (see page 182), so they created SCIgen. SCIgen is a nifty piece of software that seamlessly weaves together gobbledegook into grammatical sentences and presents it in a familiar format, ready to be submitted to conferences. They generated a couple of papers, stuck their names on them, and sent them off to the World Multiconference on Systemics, Cybernetics and Informatics (WMSCI), a conference that Maxwell Krohn, one of the creators of SCIgen, says was notorious for 'being spammy and having loose standards'.[59]

Their paper 'Rooter: A Methodology for the Typical Unification of Access Points and Redundancy' was immediately accepted as a non-reviewed paper (because reviews had not been received by the deadline).* They accepted in style, with an email containing no less than three smileys.[60]

The three planned to attend the conference, but the organisers eventually got wind of what was going on and withdrew their invitation amidst growing international media attention. The organisers sent the authors a four-page letter that one professor described as 'a mind-

---

* The other paper was rejected, though no reasons were given. When the authors asked if they might see the peer review comments, they got a rambling response from the organisers. Citing studies regarding the prevalence of such practices in journals, they said: 'If this kind of complexity seems not to be always feasible for journals, it will have less probability of being feasible for a conference. In our case we are very sorry we are not finding it feasible.' So, 'no', then.

boggling, rambling rationalization, written in full-bore buzzwordia academic'.[61] The students were not easily deterred. Capitalising on the building momentum, they raised $2,500 in just 72 hours to travel to Orlando (not bad given that this was before the golden era of viral videos and crowdfunding). They rented out a room at the same hotel as the conference and proceeded to hold their own session, which consisted of randomly generated talks by academics with fake names, fake business cards, and fake moustaches.

SCIgen is free to download, and has taken on a life of its own as scientists have used it to have a bit of fun and further expose poor publishing practices in the process. There are plenty of examples, though one published SCIgen paper stands out for its unusual author list: Marge Simpson, Kim Jong Fun, and Edna Krabappel. Alex Smolyanitsky, a researcher at the US National Institute of Standards and Technology, refused to pay the *Aperito Journal of Nanoscience Technology* $459 to publish it, but they did anyway. The paper remains freely available on the journal's website.[62] In addition to the atypical author list, SCIgen churned out some preposterous passages, such as:

> *Is it possible to justify the great pains we took in our implementation? No. With these considerations in mind, we ran four novel experiments ... We deployed 98 Motorola bag telephones across the Internet-2 network, and tested our flipflop gates accordingly.*

In December 2013, computer scientist Navin Kabra had his bogus paper, 'Use of Cloud-Computing and Social Media to Determine Box Office Performance',[63] accepted to a conference. He was trying to highlight the pitfalls of policies at his university that forced students to publish, usually in the proceedings of low- or no-standard conferences. In the introduction, Kabra (claiming to be from the 'Sokal Institute of Technology') explicitly warns the reader that what follows is meaningless drivel:

*You should read any paragraph that starts with the first 4
words in bold and italics – those have been written by the
author in painstaking detail. However, if a paragraph does
not start with bold and italics, feel free to skip it because it is
gibberish auto-generated by the good folks at SCIGen.*

The paper occasionally pretends to discuss the purported topic, including discussion of UIB and AAF algorithms (later revealed to be 'Use IMDB.com via a Browser' and 'Ask a Friend' respectively). The paper includes nineteen lines about the 1970s Bollywood film *Sholay*, and another nineteen taken directly from the 1992 Hollywood film *My Cousin Vinny*. Following one remarkably nonsensical passage, the paper states: 'The motivated reader is encouraged to not read too much into the previous paragraph, because it was copy-pasted from a random document on the internet.' The organisers claimed that the paper was one of only 60 submissions accepted of the 130 received and that all papers were double-blind reviewed by international experts.[64]

In a similar bid to expose junk journals piggybacking on university publication requirements, Mikhail Gelfand from the Russian Academy of Sciences translated the original SCIgen paper into Russian and submitted it to the Russian language *Journal of Scientific Publications of Aspirants and Doctorants*. The journal accepted Gelfand's paper and charged 4,000 Rubles (£40) for publication. However, his protest hit the mark and the government revoked their accreditation of the journal two weeks later.

French researcher Cyril Labbé of Joseph Fourier University in Grenoble catalogued SCIgen papers that had made it into over thirty published conference proceedings between 2008–13. His work revealed that 16 nonsense papers had been published by the publishing giant Springer, while the US Institute of Electrical and Electronic Engineers (IEEE) had published over a hundred. Labbé privately informed the publishers, who subsequently took steps to remove the offending papers. Labbé has since developed a program to spot fake papers by comparing an uploaded manuscript to papers known to have been generated using SCIgen.*[65]

One of the creators of SCIgen notes that Labbé's work revealed just how deep this problem runs, stating that he is proud of the program and the fact that it continues to expose weaknesses in the world of science. 'I'm psyched,' he said in an interview with the *Guardian*. 'It's so great. These papers are so funny; you read them and can't help but laugh. They are total bullshit. And I don't see this going away.'[66]

In a systematic study of sketchy publishing practices, *Science* correspondent John Bohannon published 'Who's Afraid of Peer Review?', an investigation into the peer-review processes among fee-charging, open access journals. Between January and August 2013, he submitted a fake scientific paper to 304 journals. The paper was considerably more plausible than anything SCIgen spews out, but was nevertheless written with such serious and self-evident scientific flaws that editors and peer reviewers should have summarily rejected it.† Nonetheless, 60% of the journals accepted it. *The Economist* dubbed it 'Science's Sokal moment'.[67]

Bohannon used Beall's List of predatory publishers and the Directory of Open Access Journals to build a list of 304 targets.‡ Journals accepting the paper were not only the usual suspects, but also included those from big names like Elsevier, Sage, Wolters Kluwer, and several universities. India emerged as the largest base for such publications, with 64 publishers – over 90% of them – accepting the paper. The US came in second with

---

* The program is freely available, leaving publishers and conference organisers with no excuse for accepting such papers in the future.

† Bohannon programmed a 'scientific version of Mad Libs' to vary the paper he sent to each journal. The papers all described the discovery of a new cancer drug extracted from a species of lichen, following the template: Molecule X from lichen species Y inhibits the growth of cancer cell Z. A database was set up to substitute X, Y, and Z for real molecules, lichens, and cancer cells. The data provided did not support the claimed conclusion and had obvious flaws.

‡ I.e. Fee charging, English language, open access publishers with at least one medical, biological, or chemical journal (in total, 167 from the DOAJ, 121 from Beall's list, and 16 that appeared in both).

29 publishers accepting the paper and 26 rejecting it. Nigeria was the largest African offender, with all of the journals there accepting the paper.

Because Bohannon's exposé focused only on open access publishers, it quickly became part of the polarising debate around the evolution and future of scientific publishing, with open access advocate Michael Eisen commenting that accusing the open access model of enabling internet scamming is 'like saying that the problem with the international finance system is that it enables Nigerian wire transfer scams.'*68

As is often the case in academia (and in basically all my romantic relationships to date), we agree on much of the substance, but argue vehemently about specifics and semantics. Here's my summary: let's not condemn all open access journals because of a few unscrupulous actors, but let's also be careful not to shoot the messenger when studies call out bad practices.

Unsubscribe

David Mazières and Eddie Kohler submitted a paper entitled 'Get me off Your Fucking Mailing List' to WMSCI 2005 (the same conference that accepted the original SCIgen paper). The paper consists of the title sentence, repeated over and over.

---

*   On a somewhat unrelated note, both Eisen and Bohannon are super-cool scientists and a credit to the academy. Eisen is a renowned computational biologist, a co-founder of PLOS, and has announced his intention to run for the US Senate in 2018 as an Independent science-focused candidate. He also has a keen sense of humour. The biographies of staff on his lab's homepage include important details such as which hand they use to pipette, the person's favourite statistical test, and their p-value (Eisen's is 1.72414e-06). Eisen produced an awesome 'You Have Died of Peer Review' t-shirt and his blog includes a recipe for a Vegan Thanksgiving Picnic Pie that looks absolutely incredible. Bohannon is a great science writer and has an impressive track record as a journalist. After embedding in Afghanistan in 2010, he convinced the US military to voluntarily release civilian casualty data, and he received a Reuters environmental journalism award in 2006 for his reporting on the water crisis in Gaza. He also runs the annual 'Dance Your PhD' contest, and wrote a paper entitled 'Can People Distinguish Pâté from Dog Food?' (following which he convinced US talk show host Stephen Colbert to eat cat food live on air).

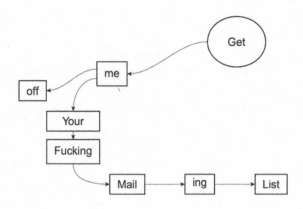

Figure 10: Get me off Your Fucking Mailing List

The paper didn't make it, but it got a second chance in 2014 when Peter Vamplew of Federation University Australia forwarded it to the *International Journal of Advanced Computer Technology* as a retort to their spam email.[69] The paper was then 'reviewed', rated as 'excellent', and accepted for publication (though the reviewer did ask Vamplew to update the references). Vamplew declined to pay the $150 article-processing fee and so the paper was ultimately not published.

It is not known whether he was removed from the mailing list.

INTERVIEW

MALE, MAD AND MUDDLE-HEADED ACADEMICS IN KIDS' BOOKS

Melissa Terras is the Director of the Centre for Digital Humanities and a Professor of Digital Humanities at University College London. She is also an expert on the portrayal of academics in kids' books, having analysed almost *300* titles.

How did you end up with a library of kids' books featuring academics?

I'm keen to share my love of books with my three kids, so we read a lot. One week I came across two different professors in children's books in quick succession. I thought it'd be a fun project to see how academics are portrayed. This turned out to be both an excuse to buy more books and a way to explain to my kids what Mummy actually does.

How do you find the books?

For four years I searched for new finds in the little bits of spare time I get throughout the day. Often academics appearing in books are not named in the title and therefore don't turn up easily via electronic searches, so I also began to obsessively search the shelves at our local library and friends' houses, and waiting rooms at doctors and dentists.

Fortunately I don't always have to do the digging myself as librarians from all over the world send me leads. People occasionally sidle up to me after a guest lecture and whisper, 'I have a good professor for you ...'

What's the oldest example you've found?

The earliest goes all the way back to 1850 – a time when the world had far fewer higher education institutions. Indeed, given the exponential growth of universities and the publication of 1.6m English language kids' books in the intervening 150 years, 281 academics seems disappointingly low.

What are the academics in children's books like?

I usually summarise them as 'male, mad and muddle-headed'. There is a lot of lazy stereotyping. Academics tend to be either crazy evil egotists (such as 'Mad Professor Erasmus', the maddest evil professor in the world) or kindly, but baffled – obsessive eggheads who don't quite function normally.

They are mostly white and male.♀ Across all the books, there are only 26 women and 3 minorities represented. Only one character is both. Professor Wiseman in the recent Curious George books is described as 'American, likely with Indian ancestry' (though in the earlier books, Wiseman was a white male).

There is some surprising variety though – Professor Peabody is a vegetable.

Professor Peabody aside, is this a fair depiction?

These depictions have their roots in public perception (and fear) of science, particularly after the Second World War, as well as broader societal trends of anti-intellectualism and structural misogyny. Looking at professors in children's books holds a mirror up to the academy itself: can we really blame children's books for not being more diverse if the academy itself is stale, male, and pale?

What kinds of stories are the books telling?

The theme of the stories tends to be 'academic is out of touch with how the world works, with hilarious consequences' in the case of professors, or 'is evil and wants to take over the world, but is thwarted by our plucky hero (never heroine)' in the case of doctors.

Any favourite characters?

The eccentric Professor Blabbermouth, and Dr Hatchett, who, having failed to find an academic job after her English Literature PhD, now teaches primary school pirates. The Boffin Boy series, written by David Orme for older kids that are struggling with reading, has proven consistently popular with the whole family. It features the stereotypically boring Professor Mudweed, as well as our only evil female, Doctor Daphne.

Your work here is unlikely to ever be finished. What's next?

Cambridge University Press will publish my book very soon – *Male, Mad and Muddleheaded: The Representation of Academics in Children's Illustrated Books*. I should think about tackling non-illustrated texts for older children next.

I can't wait to read it. Any plans to write your own kids' book?

I'm thinking about it. I would love to write a kids' book, but I can't draw and I've never written for kids! I'd need a partner.

Finally, a favourite quote:

> *'Professor Blabbermouth was as bright as buttons. There*
> *was no doubt about it. She had enough university degrees*
> *to paper her toilet walls. Some people said she was a genius.*
> *Some people said she was a nutter. It was all a matter*
> *of opinion ... All those brains and nothing to use them*
> *on made her do rather ... eccentric things. Like cycling*
> *backwards to the shop in the belief it saved time. Or for*
> *a complete week never using the letter 'e' whn spaking to*
> *popl. She never explained her reasons for this. And nobody*
> *thought to ask.'*

Figure 11: The Professor's Lecture[70]

OBSCURE INTERLUDE

---∞∞∞---

BEARDS

Academia has long been the bastion of beards, and now they are making a hipster-fuelled comeback outside the ivory tower too.

As a result, a 2014 study in *Biology Letters* suggested that we are fast approaching 'peak beard', the point at which beards are so common that they become undesirable from an evolutionary perspective.[1] The researchers showed participants a range of pictures of faces, manipulating the frequency of beards, and then measured preference for four levels of beardedness. Both women and men found heavy stubble and full beards more attractive when presented with a set of faces in which beards were rare. Likewise, clean-shaven faces were least attractive when such faces were common, and more attractive when rare.

Such peaks are apparently cyclical. A previous review of facial hair styles found that sideburns peaked in 1853, moustaches in 1877 and beards in 1892.[2] Moustaches subsequently had a renaissance, before peaking again from 1917 to 1919. The study also noted a positive correlation between the prevalence of beards in men and the average width of women's skirts – as beards become more common, skirt widths increase.

In 'Beards: An Archaeological and Historical Overview', the author notes:[3]

Beards have been ascribed various symbolic attributes, such as sexual virility, wisdom and high social status, but conversely barbarism, eccentricity and Satanism.

Studies have indeed returned mixed results. In one study, full beards rated highest for parenting ability and healthiness,[4] while in another, bearded men with an aggressive facial expression were perceived as being significantly more aggressive than the same men when clean-shaven.[5] One study even considered whether a woman's menstrual cycle affects their perception of beards.[*6]

While a beard might provide a small amount of sun protection,[7] there are concerns that bearded scientists could inadvertently harbour dangerous microorganisms or chemicals in their face fur. A 1967 paper published in *Applied Microbiology,* aimed to evaluate the risks.[8] What ensues is a bizarre study that involved spraying pathogens on academics' beards (73-day-old beards, to be precise), washing their faces, then collecting some beard dust to see if the pathogens were still present. The paper also documents a second study testing the pathogen-infested beards on chicks using an ultra-creepy human-head mannequin.

After much contamination, washing, and the needless death of a handful of sentient beings, the authors find that a beard would only pose a risk following a 'recognizable microbiological accident with a persistent highly infectious microorganism', or if the wearer was 'engaged in a repetitious operation that aerosolized a significant number of organisms'.

Figure 12: Chickens exposed to natural hair beard on mannequin

---

* Not really: 'preferences vary only subtly with respect to hormonal, reproductive, and relationship status'.

WRITING

Planning to write is not writing. Outlining, researching, talking to people about what you're doing, none of that is writing. Writing is writing.
E. L. Doctorow

I love deadlines.
I love the whooshing noise they make as they go by.
Douglas Adams

There is nothing to writing.
All you do is sit down at a typewriter and bleed.*
Ernest Hemingway

---

* Modern academic writing tends to be more about sitting down at a laptop and despairing at your poor life choices.

A PASSAGE REGARDING SUCCINCTNESS AND THE EXIGENCIES OF PROACTIVELY COUNTERACTING SESQUIPEDALIANISM IN ACADEMIC COMPOSITION[*][1]

Academics are not known for their ~~concise writing~~ concision. We have a reputation for droning on in language strewn with jargon and unnecessarily long words. Yet on occasion the rare brevity seen earlier with the one-word abstracts can be observed in academic papers.

A 2003 paper, 'Higher taxa: Reply to Cartmill', consists of two words: 'Enough already.'[†] This was the final shot fired in a year-long back-and-forth between Ian Tattersall, curator emeritus at the American Museum of Natural History, and Boston University professor Matt Cartmill. Cartmill kicked off with his paper 'Primate Origins, Human Origins, and the End of Higher Taxa', to which Tattersall replied with 'Higher Taxa: An Alternate Perspective'. Cartmill hit back with 'The End of Higher Taxa: A Reply to Tattersall', before Tattersall finally declared that he'd had enough already. The two (who I believe are otherwise good friends) have been battling each other since the 1980s over various arcane details of systematics.[‡]

Similarly terse is a sarcastic paper regarding the use of the term 'chemical-free'.[2] The authors first declare that their aim is to describe

---

[*] In his paper, 'Consequences of Erudite Vernacular Utilized Irrespective of Necessity: Problems with Using Long Words Needlessly', Daniel Oppenheimer assesses the hypothesis that using long words makes you seem smarter (they don't).

[†] The keywords to the paper are: enough; already.

[‡] That is, the study of the diversification of living organisms. If I understand correctly, and there is every chance that I do not, Cartmill questions why tiny differences are sometimes taken to separate certain animals into different species and families, while others aren't, whereas Tattersall sees this position as an attack on the field of systematics itself, it being essential to document even the tiniest of changes and classify species accordingly.

Figure 13: The writing process

all consumer products that are appropriately labelled as 'chemical-free', then they explain that this is a misnomer because everything contains chemicals, and then follow up with two blank pages. The only other text is a footnote at the end of the paper declaring that the authors have no competing financial interests, but 'would have short-sold "Rubber Ducky Sunscreen" on principle if it was publicly traded' (according to its website, Rubber Ducky is a '100% Chemical-Free' sunscreen).

Mathematics is the field with the longest history of shortest papers. Euler's conjecture – a theory proposed by Leonhard Euler in 1769 – survived unchallenged for 200 years, until two mathematicians unceremoniously debunked it in 1966 with just two short sentences printed in the *Bulletin of the American Mathematical Society*.[*3] Others have matched the two-sentence record, though none have shattered any 200-year-old conjectures in the process.

---

* The entire article reads: 'A direct search on the CDC 6600 yielded $27^{-195}+84^{-195}+110^{-195}+133^{-195}=144^{-195}$ as the smallest instance in which four fifth powers sum to

a fifth power. This is a counter-example to a conjecture by Euler that at least n nth powers are required to sum to an nth power, n>2.' The *CDC 6600* used by the authors is generally considered to be the first successful supercomputer. It was the world's fastest computer at the time, outperforming the closest competitor, the *IBM 7030 Stretch*, by a factor of three. It remained the fastest in the world until 1969 when it was outpaced by its successor, the *CDC 7600*. IBM was concerned that it was being beaten by CDC, a much smaller company, leading IBM CEO Thomas J. Watson to write a memo to staff: 'I understand that in the laboratory developing the system there are only 34 people including the janitor. Of these, 14 are engineers and 4 are programmers ... Contrasting this modest effort with our vast development activities, I fail to understand why we have lost our industry leadership position by letting someone else offer the world's most powerful computer.' The electrical engineer that created the CDC 6600, Seymour Cray (often called the 'father of supercomputing') responded: 'It seems like Mr. Watson has answered his own question.' In 2011, Michio Kaku observed that 'your cell phone has more computer power than all of NASA back in 1969, when it placed two astronauts on the moon'.[4] By the same token, that sleek slab of glass and plastic in your pocket (that you mostly use to crush candy and fling birds at pigs) has far greater processing power that the CDC 6600 (the maximum speed of the CDC6600 was 3 megaFLOPS (millions of floating point operations per second) while the iPhone 5's graphics processor alone can hit 76 gigaFLOPS (billions of floating point operations per second) – 25,000 times more).[5] The first CDC 6600 was delivered to CERN in Geneva in 1965, where it was used to analyse the 2–3 million photographs of bubble chamber tracks that their experiments were producing each year (a bubble chamber is a piece of apparatus used in physics to 'see' particles by photographing the tracks of bubbles left by ionising particles as they move through a superheated transparent liquid (usually liquid hydrogen) (CERN's website has lots of cool photos)).[6] The bubble chamber was invented in 1952 by Donald Glaser; he was awarded the Nobel Prize for Physics in 1960. Legend had it that Glaser's inspiration for the bubble chamber came from the bubbles in a glass of beer. In a 2006 talk he corrected this story, noting that while beer was not the inspiration for the bubble chamber, he did experiment with using beer as the liquid to fill early prototypes.[7] Beer has nonetheless been used to demonstrate the exponential decay law,[8] inspired a theory on the impact of the moon's phases on sleep and diagrams of particles that look like penguins (see pages 119 and 205), and fuelled many hours of writing for this book. All that is in spite of the fact that a study in the Czech Republic hypothesised that beer consumption lowers academic productivity.[9] On the topic of beer, the medical literature contains a report of a man with so-called 'Auto-Brewery Syndrome':[10] the presence of *Saccharomyces cerevisiae* in the man's gut caused the spontaneous brewing of alcohol nearly 24 hours after the ingestion of sugar, meaning that he was frequently intoxicated despite not having touched a drop. But I digress ...

This trend for short maths papers culminates with the paper 'Can n2 + 1 unit equilateral triangles cover an equilateral triangle of side > n, say n + ε?'. The body of the paper consists solely of the text 'n2 + 2 can', followed by two diagrams. Professor Alexander Soifer[*] recounts that *American Mathematical Monthly* was taken aback by his article.[11] Two days after submission, an editorial assistant acknowledged receipt of the paper, but stated that it 'is a bit too short to be a good *Monthly* article...A line or two of explanation would really help.'[12] Soifer consulted with his co-author, John Conway, over coffee. His equally concise response was: 'Do not give up too easily.'

Soifer fired back the same day to make his case:

> *I respectfully disagree that a short paper in general – and this paper in particular – merely due to its size must be 'a bit too short to be a good Monthly article'. Is there a connection between quantity and quality?... We have posed a fine (in our opinion) open problem and reported two distinct 'behold-style' proofs of our advance on this problem. What else is there to explain?*

Less than a week later they received a response from Editor-in-Chief Bruce Palka offering to publish the paper in a box on a page that would have otherwise contained a lot of blank space. The authors accepted and the paper was published.[†]

Nanopublications
While the preceding examples are mostly concise for comic effect, it does

---

[*] Soifer has an Erdős number of 1, as Erdős was his PhD supervisor (See page 148). Unrelated: Soifer teaches an uncommon combination of math, art, and film history at the University of Colorado.

[†] Insisting they must have the last word, the publisher moved the original title to the body of the paper and added a more substantive title without consulting the authors.

seem increasingly clear that the shortened attention spans of the social media era will make such brevity increasingly necessary. In any case, condensing research results into digestible chunks is a reasonable response to the overwhelming quantity of literature that academics have to sift through.

Although the momentum to develop the world's first Twitter-only journal appears to have stalled,[13] the journal *Tiny Transactions on Computer Science* (*TinyToCS*) has begun in earnest, publishing computer science research of 140 characters or less.* These handy snippets, dubbed 'nanopublications', are the 'smallest unit of publishable information: an assertion about anything that can be uniquely identified and attributed to its author.'[14] *TinyToCS* published a nanopublication about nanopublications, which serves simultaneously as both an explanation and an example of the format. The entire paper reads:[15]

> *The nanopublication model incentivizes rapid, citable data dissemination, interoperability, semantic reasoning, and knowledge discovery.*

WRITING IS DIFFIC

If Tattersall's two words is two too many, or if nanopublications still seem too long, a series of papers on 'Writer's Block' may be the antidote.† In 1974 psychologist Dennis Upper 'wrote' an academic paper containing precisely no words, entitled 'The unsuccessful self-treatment of a case of "writer's

---

* In reality, nanopublications aren't quite as diminutive as their name might suggest – though the body of the article is always a single statement, it is generally accompanied by a much longer 'background' section resembling a traditional abstract.

† The Google Books Ngram Viewer suggests that this term only came into usage in 1945. It had a period of minimal usage until about 1965, when it really started to take off. Usage shot up until about 1987, when there was a sudden dip. Neologisms are central to academia – and to nonsense.

block"'.*[16] The enthusiastic peer reviewer stated that they examined the manuscript with lemon juice and X-rays and did not find a single flaw, concluding that the paper should be published without revision. ('Surely we can find a place for this paper in the journal – perhaps on the edge of a blank page.')

In 1983, Geoffrey Molloy published a replication in which he also failed to put pen to paper,[17] though a year later Bruce Herman advanced the literature ever so slightly in his 'partial failure to replicate':[18]

> *Self-treatment of 'writer's block', while generally reported to*
> *be unsuccessful (Molloy, 1983; Upper, 1974), may not be*
> *entirely without merit. I say this becau*

Herman notes that the paper was supported by a grant from the American Institute of Communicative Disorders† and that portions of the paper were presented at the First Annual Convention of the International Association to Combat Writer's Block (presided over by Isaac Asimov).

A group of authors then published their unsuccessful group-treatment of a case of writer's block,[19] in which 'a regime of weekly 1-hr. sessions over a 2-yr. period was ineffective in remediating writer's block in any of the five participants.' The group conducted a follow-up assessment a decade later.[20] Treatment had continued to be unsuccessful, which the authors postulate might be due to '(a) second author's relocation to another university, and (b) apparent inability of the other original participants to respond to posthumous treatment.'

Another decade passed before Didden et al. published 'A Multisite

---

*   A footnote to the title reads 'Portions of this paper were not presented at the 81st Annual American Psychological Association Convention'. To me this suggests that portions of the paper *were* presented at the Convention, which would presumably involve a 'presentation' consisting entirely of silence, a la John Cage's 4'33".

†   I attempted to confirm this, but the representative from the *American Institute of Communicative Disorders* was incredibly rude to me and refused to comment.

Cross-Cultural Replication',[21] i.e. a blank page written by several authors on different continents. This time the authors state that the article was supported by a $2.50 grant from the first author's personal funds, and that they hope to submit the paper to the 'next international conference in St Tropez'. Upper's blankness had stood the test of time and the reviewer was once again enthusiastic, commending its 'awe-inspiring brevity'.

The latest in this long line of papers came in 2014 with Mclean and Thomas's meta-analysis, which concludes: 'Group-treatments tend to be slightly more unsuccessful than self-treatments.'[22]

This research isn't getting us any closer to a cure, yet in 1925 Hugo Gernsback, one of the pioneers of science fiction, may have already invented it. In *Science and Invention* magazine, he showcased one of his bizarre creations, 'The Isolator'. The cumbersome contraption, which resembles a cross between a giant gas mask and an old-school diving helmet, was intended to encourage focus and concentration by eliminating external sensory stimuli. The helmet completely blocked out sound, limited vision to a tiny horizontal slit, and supplied the writer with pure oxygen.

I'll stick to the library.

Figure 14: The Isolator

TRIPE

Getting the words flowing is a difficult, sometimes seemingly insurmountable, first step in any writing project, but the real challenge is writing concisely and comprehensibly.

Social Text, target of Alan Sokal, is known for publishing some particularly perplexing articles. 'S'More Inequality – The Neoliberal Marshmallow and the Corporate Reform of Education', singled out by Marc Abrahams, is one such paper.[23] Keen for a challenge, I had a go at reading it. It was bloody difficult. One scholar posted on Twitter that it was a 'fascinating read', (though sarcasm is notoriously hard to detect in written form).[*][24]

Here is an extract of the abstract:

> *The marshmallow test is more than a handy synecdoche for the cold new logic behind shrinking public services and the burgeoning apparatus of surveillance and accountability. It also shows how the sciences of the soul can be deployed to create the person they purport to describe, by willing political transformation.*

Quite.

If this paper dances gaily on the fringes of comprehensibility, another of Abrahams's collected oddities, a paper published in *Qualitative Inquiry* entitled 'Welcome to My Brain', is baffling beyond belief.[25] Reading it is akin to being inside a migraine, and one puzzled scientist asked his colleagues to read it so he could be sure he hadn't had a stroke.

The keywords for the paper include 'de/re/subjective twisted/ing

---

[*] A reversed question mark (؟) appears to be the frontrunner solution to this problem. It was proposed by English printer Henry Denham in the 1580s and used by Marcellin Jobard and French poet Alcanter de Brahm during the 19th century. Ethiopic languages already use a mark to denote sarcasm, *Temherte Slaqî*, which is indistinguishable from an inverted exclamation mark (¡).

brain de/re/construction', and a sizeable chunk of the text is dedicated to telling the reader what the paper is about (with little success). The abstract reads:

This is about developing recursive, intrinsic, self-reflexive as de-and/or resubjective always evolving living research designs. It is about learning and memory cognition and experiment poetic/creative pedagogical science establishing a view of students ultimately me as subjects of will (not) gaining from disorder and noise: Antifragile and antifragility and pedagogy as movements in/through place/ space . . . I use knitting the Möbius strip and other art/ math hyperbolic knitted and crocheted objects to illustrate nonbinary . . . perhaps. Generally; this is about asking how-questions more than what-questions.

Also seeking reassurance that I hadn't suffered a stroke, I read this over the phone to a close friend, comedian and confidant Haydn Griffith-Jones. He proffered that maybe it only *sounded* complex, but in reality was quite simple. By way of example he recounted that he'd recently been perusing the wares of an online sex toy retailer and had found the 'double penetrator strap-on vibrating rabbit cock ring' to be considerably less complex and intimidating than its name would suggest. I looked it up and can confirm that both are every bit as complex (and ridiculous) as they sound.

For reasons that are never elucidated, the author makes repeated reference to Möbius strips and some bloke called John. In one singularly dense paragraph the author begins, 'Knitting John, John knitting. Knitting John Möbius. Möbius knitting John'. This is then followed by a description of how Möbius strips have been used as conveyor belts, recording tapes, and in the design of versatile electronic resistors. The passage concludes with: 'The wear and tear of my efforts. My stunts, enthusiasm knitting. My brain and doubling and John.'

At best, these papers demonstrate that unnecessarily complex language is generally unhelpful. At worst, they reinforce the preconception that academics live in cloud cuckoo land detached from reality, and prove that there is no bottom limit to the gibberish that some journals are willing to publish.

TROPES

Just as clichés plague paper titles, there are tropes and phrases used so regularly and unflinchingly in papers that they have become more or less compulsory. You must 'gratefully acknowledge' all those who helped you realise the work, your paper must 'fill a gap in the literature', further research must always be required, and, crucially, your results have to be 'significant'.

'I gratefully acknowledge ...'

In the acknowledgements section of their books and papers, researchers have thanked everyone from Rocco Siffredi (an Italian porn star) for his 'constant support',[26] to the thrash metal band Slayer for 'continued advice and inspiration',[27] to Jon Frum (a cargo cult deity).[28] Computer scientist Guillaume Cabanac thanked his daughter for helping to collect data, though in reality she was four-month-old baby sleeping by his desk,[29] while a couple of Barcelona fans working in the US managed to sneak in their home football chant, '*Visca el Barça*!'[30] Three Italian researchers went as far as including a unique section in their paper:[31]

> **Unacknowledgements:** *This work is ostensibly supported by the Italian Ministry of University and Research ... The Ministry however has not paid its dues and it is not known whether it will ever do.*

Unsurprisingly, there are a few that focus on funding. Sci-fi historian Adam Roberts wrote:[32]

> *Let me record that I am not in the least grateful to the*
> *British Arts and Humanities Research Board – A plague on*
> *their house. That this book was ever completed owes nothing*
> *to them at all.*

Evolutionary biologist Leigh Van Valen, who was 'considered unconventional even by eccentrics',[33] wrote:[34]

> *I thank the National Science Foundation for regularly*
> *rejecting my (honest) grant applications for work on real*
> *organisms, thus forcing me into theoretical work.*

An especially acerbic unacknowledgement appears in Brendan Pietsch's book *Dispensational Modernism*:[35]

> *I blame all of you. Writing this book has been an exercise in*
> *sustained suffering. The casual reader may, perhaps, exempt*
> *herself from excessive guilt, but for those of you who have played*
> *the larger role in prolonging my agonies with your encouragement*
> *and support, well . . . you know who you are, and you owe me.*

Unacknowledgements sometimes include passive-aggressive barbs aimed at those the authors feel have wronged them:

- 'We would like to thank Karla Miller for sleeping late one morning, leaving Tim and Steve a bit bored.' (They also thank one Saad Jbabdi for 'making the brains look pretty'.)[36]

- 'I thank Graham Higman for allowing the dust of Oxford to rest on my unopened manuscript for thirty months.'[37]

- 'We gratefully thank Programme National de Physique Stellaire for financial support. **We do not gratefully thank T. Appourchaux for his useless and very mean comments**.'[38] (bold is theirs)

Others explain the curious circumstances surrounding their work:

- 'Most of the paper was written during my daily commute from Vancouver to Surrey, Canada, and I would like to acknowledge TransLink Metro, Vancouver's regional transportation authority, for making the task of writing in buses and trains such an enjoyable exercise.'[39]

- 'If the book is not a success, I dedicate it to the burglars in Boulder, Colorado, who broke into our house and stole a television, two typewriters, my wife Helen's engagement ring and several pieces of cheese, somewhere about a third of the way through Chapter 8.'[40]

- '…would also like to thank the US Immigration Service under the Bush administration, whose visa background security check forced her to spend two months (following an international conference) in a third country, free of routine obligations – it was during this time that the hypothesis presented herein was initially conjectured.'[41]

- 'Research supported in part by the Federal Prison System. Opinions expressed in this paper are the author's and are not necessarily those of the Bureau of Prisons'. (The author, Chandler Davis, was serving a prison sentence for refusing to cooperate with the House Un-American Activities Committee.)*[42]

There are, of course, those who like to genuinely thank their loved ones:[43]

---

* The body created by the US government to investigate disloyalty and subversive organisations, known for its McCarthyist witch-hunts during the 1950s and 1960s.

> *This book is dedicated to my brilliant and beautiful wife*
> *without whom I would be nothing. She always comforts and*
> *consoles, never complains or interferes, asks nothing, and*
> *endures all. She also writes my dedications.*

Caleb Brown from the Royal Tyrrell Museum helps us end this section on a positive note. His acknowledgements in a *Cell* paper describing a new dinosaur read:[44]

> *C.M.B. would specifically like to highlight the ongoing and*
> *unwavering support of Lorna O'Brien. Lorna, will you*
> *marry me?*

She said yes.

Gap in the literature

Academics often say that their much-needed paper fills a gap in the literature, but it would be more accurate to say that they create a much-needed gap in the literature.[*][45] This is, in reality, what most papers are doing – carving out a tiny niche to justify their existence.

There is a gap in the literature for everything. There is a gap in the literature for dressing up as a polar bear to try and scare reindeer.[46] There is a gap in the literature for modelling avalanches by chucking 300,000 ping-pong balls down a ski jump.[47] There is a gap in the literature for looking at bareback sex through the lens of queer legal theory.[48] There is a gap in the literature for analysing *Fifty Shades of Grey* using the writings of obscure ancient Greek philosophers.[49] There is most definitely a gap in

---

* I thought I was being clever, but I am not the first to make this joke. The phrase first appeared around 1960 in a review for *Mathematical Reviews*, wherein Lee Neuwirth, then an instructor at Princeton, began a review of an article by Hale Trotter with the sentence. The phrasing was unintentional (or at least subconscious), such that when Neuwirth showed the review to his colleague Ralph Fox he 'roared with laughter'.

the literature for you to justify whatever crazy thing it is that you want to research.*

Table 3: More super-specific gaps in the literature

Gap	Conclusion
Analysing the body composition of Spanish football referees[50]	The average Spanish referee is 32 years old, weighs 72.3 kilograms, and is 1.79 metres tall.
Searching the internet for evidence of time travellers[51]	No time travellers were discovered.
Working out why people hated Clippy, the Microsoft Word assistant[52]	Clippy was apparently built to invoke rage: it breaks basic rules of etiquette, unduly disturbs users, and doesn't even provide a helpful service.
Calculating how much gravity needs to weaken before we can walk on water[53]	If the moon had water, a person could run on it using small fins.
Using bacteria from baby poo to make fermented sausages[54]	It's theoretically possible, but literally nobody is going to buy them.

More research required

A sentence or two declaring that the topic is going to need more research paves the way for the author(s) to do said research themselves in the future. Indeed, academics that are truly on top of their research agenda often have the next paper in the pipeline. There are lots of ways to make this declaration, such as:

---

* Writing this, I am reminded of the words of Felipe Andres Coronel (aka rapper *Immortal Technique*), who, in lamenting the lack of diversity and variety in commercial hip-hop, says: 'There is a market for everything man. There is a market for pet psychologists ... For nipple rings, for river dancing, for chocolate-covered roaches ...' Take it from *Tech*, there is always a gap in the literature.

- 'We can only see a short distance ahead but we can see plenty there that needs to be done.' (Turing on artificial intelligence).[55]

- 'Even if it is correct, it is clear from what we have said that much remains to be discovered...' (Watson & Crick on the structure of DNA).[56]

- 'It needs not only new applications, but also improvements, further development, and plenty of fresh energy.' (Mendeleev on the periodic table).[57]

At the end of Paul Krugman's paper on interstellar trade, he concludes:

I have not even touched on the fascinating possibilities of interstellar finance, where spot and forward exchange markets will have to be supplemented by conditional present markets. Those of us working in this field are still a small band, but we know that the Force is with us.

The blunt parting shot of a paper written way back in 1900 was:[58]

This work will be continued and I wish to reserve the field for myself.

Significance

Everybody wants their work to be important, and in academia importance means statistical significance. Enter the p-value. P-values are used to denote the significance of a given result, and p-value of less than 0.05 (i.e. the outcome would happen by chance no more than 5% of the time) has somewhat arbitrarily emerged as the benchmark for significance.

As a result, academics do everything they can to make sure their findings pass this threshold. When a p-value remains stubbornly higher than 0.05, academics are reluctant to tell the truth, and instead have come up with myriad ways to say that they just missed the mark.

Statistician Matthew Hankins has compiled a list of 500 ways that academics have minced their words when describing the significance of their results.* Here are 13[†] examples of authors keen to honestly reassure you that they only just very narrowly missed out on the traditionally accepted threshold for statistical significance by the most vanishingly small of margins.

Table 4: Selected p-value workarounds

Hypothesis	Quote	p-value
The Peters et al. Delusions Inventory is a better test than the General Health Questionnaire at discriminating patients with a mental disorder with psychotic features from putatively healthy people.	'A barely detectable, statistically significant difference'[59]	0.073
Consumption of South American psychoactive beverage Ayahuasca increases systolic blood pressure.	'A robust trend toward significance'[60]	0.0503
Difference between the sexes in the skeletal development of the hands and wrists in Finnish children.	'Barely escapes being statistically significant at the 5% risk level'[61]	$0.1 > p > 0.05$

---

* Specifically he sought out papers from peer-reviewed journal articles in which: (a) the authors set themselves the threshold of 0.05 for significance, (b) failed to achieve that threshold value, and (c) described it in such a way as to make it seem more interesting.

† I read somewhere that providing an uneven number of items in a list increases the intrigue.

Migration of Immunoglobulin A*-bearing lymphocytes† into saliva.	'Bordered on, but was not less than, the accepted level of significance'[62]	>0.05
Women are more likely than men to oppose immigrants from richer countries and support immigration from poorer countries.	'Only flirting with conventional levels of significance'[63]	>0.1
Something to do with shipping routes.	'Hovers on the brink of significance'[64]	0.055
Something to do with oxygen consumption by tropical butterflies.	'Just tottering on the brink of significance at the 0.05 level'[65]	Not specified
Higher UV absorption in water reduces toxicity of silver to the freshwater crustacean *Daphnia magna*.‡	'Narrowly eluded statistical significance'[66]	0.0789
The creatine phosphate, acid-soluble and total phosphorus contents of the skeletal muscle of a rat drops after one hour in a pressure chamber at a 'height' of about 10,000m.	'Not absolutely significant but very probably so'[67]	>0.05
Increase in vegetable consumption during pregnancy reduces mercury levels in maternal blood, cord blood, and meconium.§	'Not very definitely significant from the statistical point of view, it was at the boundary of significance'[68]	0.08

---

* An antibody that plays a critical role in mucosal immunity.
† A type of white blood cell.
‡ Daphnia magna, a type of water flea, is commonly used as a laboratory animal for testing ecotoxicity because they are small, easy to raise, and produce genetically linked offspring through asexual reproduction.
§ The content of a baby's earliest bowel movements.

Community mental health teams in rural communities in England are less well integrated than teams that served urban or mixed populations.	'On the very fringes of significance'[69]	0.099
The effect of emotional conflict on attention allocation.	'Tantalisingly close to significance'[70]	0.104
	'Did not reach the traditional level of significant, but it resides on the edge of significance'[71]	0.1

MIND YOUR LANGUAGE

Despite their love of copy-and-paste tropes, academics can sometimes surprise with evocative language or offbeat style:

- M. N. Huxley compares mathematics to an orchestra: 'Poisson summation is the tuba: very deep, but ridiculous when used too much.'[72]

- Fellow mathematician Peter Johnstone cites Milne (1926), i.e. *Winnie the Pooh*, in describing the proof of a theorem as: 'A fairly straightforward Woozle-hunt.'*[73]

---

* While it may be a cute phrase, a Woozle-hunt would not be a straightforward, or useful, proof technique. You may recall that in the world of Winnie the Pooh, a Woozle-hunt involves going round in circles for an extended period of time, ultimately ending without the capture of any Woozles. Achieving proof by Woozle-hunt would therefore be a considerable achievement. Incidentally, the 'Woozle effect' is a term sometimes used to describe evidence by citation, i.e. when frequent citation of previous publications that lack evidence misleads readers into thinking that there is evidence.

- David A. Cox and Steven Zucker created an algorithm called the Cox–Zucker Machine.[74]

- To excuse his supposedly poor English, Hermann Weyl writes, 'The gods have imposed upon my writing the yoke of a foreign language that was not sung at my cradle.'[75]

- A paper on super-massive black holes remixes the epigraph to *The Lord of the Rings*, replacing both 'ring' and 'Mordor' with 'Sérsic' (a mathematical function that describes how the intensity of a galaxy varies with distance from its centre).[76]

> *Three Sérsics for the Elven-kings under the sky,*
> *Seven for the Dwarf-lords in their halls of stone,*
> *Nine for Mortal Men doomed to die,*
> *One for the Dark Lord on his dark throne,*
> *In the Land of Galaxies where the Shadows lie,*
> *One Sérsic for strong residuals,*
> *One Sérsic to fiat them,*
> *Three Sérsics to bring them all and in the darkness bind them*
> *In the Land of Sérsic-fits where the Shadows lie.*
>
> The Lord of the Sérsics, epigraph

- In 1971 the *Journal of Organic Chemistry* published a paper written entirely in iambic pentameter* (a format favoured by Shakespeare and therefore more commonly seen in poems and plays than in chemistry papers):[77]

---

* A commonly used type of metrical line in traditional English poetry and verse drama. The term describes the rhythm that the words establish in that line, which is measured in small groups of syllables called 'feet'. The word 'iambic' refers to the type of foot that is used, known as the iamb, which in English is an unstressed syllable followed by a stressed syllable. The word 'pentameter' indicates that a line has five of these 'feet'. I copied that entirely from Wikipedia. Sorry.

> Tribromobenzene isomerisations
> Are well catalysed by potassium
> Anilide in liquid ammonia.
> It was therefore of interest to see
> The effect of this base on mobility.
> Results are assembled in Table IV.

With the aim of introducing more lively language into scholarly works, the PhD Challenge was started by zombies[*] in 2010. The challenge saw fledgling scholars attempting to include a defined phrase, generally odd or obscene, into a peer-reviewed publication.

Gabriel Parent from Carnegie Mellon was the first winner, sneaking the sentence 'I smoke crack rocks' into his paper on speech-recognition systems.[78] He notes that callers can cause problems when they use language that isn't in the typically limited vocabulary of automated telephone systems. For example, a caller could yell 'I smoke crack rocks' down the phone and the computer system wouldn't have a clue what it meant. He won a box of ramen noodles and a pack of leather elbow patches for his efforts. (I was unable to confirm whether the tentative prize of an autographed photo of Paul Krugman ever came to fruition.)

The 2011 challenge was to get a paper published with at least one author with the nickname 'Dirty Old Man' or 'Crazy Cat Lady'. NYU postdoc Tom Schaul's daring exceeded expectations as he managed to co-author with ill-fated dictator Muammar 'Dirty Old Man' Gaddafi.[79] He won a Calabash professor's pipe and a copy of Strunk & White's classic *The Elements of Style*.

---

[*] Josh Bernoff's first comment when editing this book was: 'Academics overuse the passive voice. So do Brits. There is a whole lot of it.' I too dislike the passive voice, but as a British academic, I have trouble identifying it and rephrasing accordingly. Rebecca Johnson, Dean of Academics and Deputy Director of the Marine Corps War College, came up with a rule to identify passive voice: 'If you can insert "by zombies" after the verb, you have the passive voice.' This ingenious test helped me no end. However, this sentence proved difficult to reword because I was not able to identify who was behind the PhD Challenge. In the absence of a subject, I am choosing to assume that the PhD Challenge was started by zombies.

Demonstrating that a bit of humour did no harm to their career prospects, Tom now works at Google DeepMind and Gabriel went on to work at Amazon. Sadly the challenge itself is no longer running. It seemed to fizzle out around the time of the 2012 edition, meaning nobody ever named something as a 'Cleveland steamer'.*

Fuck in Nature†

Nature is full of bollocks. That is the conclusion of Stuart Cantrill (author of the *Chemical Connections* blog), who did a quantitative analysis of the use of profanity in the journal.[80] The first time bollocks got an airing in *Nature* was in 1998. The journal had published Cornelia Parker's pictures of belly button fluff (in Martin Kemp's segment on the linkages between art and science), and in a follow-up piece Kemp quoted a postgrad overheard in the Leicester University tearoom: 'What's this bollocks doing in *Nature*?'[81] Kemp's article in turn prompted a letter that begins, 'How lovely to see the word "bollocks" appearing, perhaps for the first time, in *Nature*.'

'And so,' Cantrill reflects, 'this intimately related pair of "bollocks" appeared in *Nature* within the space of two weeks.'

In addition to all the bollocks, *Nature* has featured a total of 48 'shits' (including 13 'bullshits', 1 'shit-stirrer' and 1 nano-shit), 26 'piss'-derived expressions, and a grand total of ten 'fucks' (i.e. approximately one fuck given by *Nature* every 18 years). The first fuck in *Nature* predates bollocks by almost 60 years.

A 1937 'fuck' appears in a section listing the titles of presentations, wherein one entry reads: 'Observations on the parasitism of *Sclerotinia libertiana sclerotiorum* Fuck associated with other fungi.' (The italicised name of the plant fungus, named after Karl Wilhelm Gottlieb Leopold Fuckel, was

---

* Do not google this. There are some things you can't unlearn.

† If you are easily offended by profanity, it may be best to skip over this section, which discusses the use of such words in academia. The BBC, which I shall use as my barometer for foul-mouthery, categorises just three words as 'the strongest language': cunt, motherfucker and 'fuck or its derivatives' (is 'motherfucker' not a derivative of 'fuck'?).

sometimes abbreviated in this way.) A second 'fuck' in 1985 is similarly innocent: the name of one of the authors of a cited paper is R. A. Fuck. It is only in 1989 that the word is used in its expletive form, and even then it is only in a quote in a book review – 'Oh fuck, another new phylum.'*

The most famous 'Fuck' in academia to date is a paper of that name, which explores the legal implications of the word, by Christopher Fairman, an academic at Ohio State University.[82] Fairman begins:

> 'Oh fuck. Let's just get this out of the way. You'll find no F-word, f*ck, f—k, @$!%, or other sanitized version used here.'

Fairman isn't fucking around: *Fuck* features a staggering 482 instances of the titular expletive in its extensively researched 74 pages (6.5 fucks per page). By contrast, Allen Walker Read's 1934 scholarly treatment of the word ran to 15 pages, but there is not a single use of the word itself in sight.[83]

Brian Leitner, head of the Social Sciences Research Network[†] refused to include *Fuck* in its annual calculation of law school rankings.[‡] The ranking is based on the number of downloads of the school's papers, and, reasoned Leiter, *Fuck*'s 'unusually high download count was due to its provocative title, not its scholarly content …'[84] Fairman disagreed, resulting in a protracted public exchange between the two.[85]

Fairman was far from the first to have fun with the F-word. James McCawley, a Scottish-American linguist who studied under the supervision of Naom Chomsky (and wrote a book on deciphering menus in Chinese restaurants)[86] produced a profanity-laden paper on 'English

---

* In fact, all of the remaining seven instances of the word, and its variations, are found in quotes, and only appear in news stories, features or the Books & Arts section.

† One of the biggest repositories of papers in the social sciences, recently bought by Elsevier.

‡ If the rankings had included Fairman's paper, then Ohio State (where Fairman was based) would have ranked 10th, and Emory (where he was visiting) would have ranked 8th; without Fairman's paper, neither would have been close to the top 15.

sentences without overt grammatical subject'.[87] Writing under the pseudonym Quang Phúc Đông from the fictitious South Hanoi Institute of Technology, McCawley employs colourful examples to explain why the phrase 'fuck you', and others like it, are not imperatives. For example, the sentence 'Fuck Lyndon Johnson' can be 'interpreted either as an admonition to copulate with Lyndon Johnson or as an epithet indicating disapproval of that individual but conveying no instruction to engage in sexual relations with him.' The paper features an assortment of choice phrases, like 'Describe and fuck communism' and 'Fuck complex symbols carefully'.*

I have to confess at this point that I do more than my fair share of swearing. I love bollocks, I throw out the odd shit and fuck, and, having lived my adolescence in the *American Pie* era, I generally don't take it personally if someone playfully insinuates that I have sexual relations with people's parents. But somehow the C-word still feels taboo to me (I struggle to bring myself to type it out). And I am not the only one: 'Cunt' (there, I said it) is the only one of BBC's big three yet to have an academic paper dedicated entirely to it.

Nonetheless, an accidental inclusion came in a 2007 paper published in *Chemical Communications*, which deals with the subject of copper nanotubes.[88] In a paper that refers to such nanotubes 50 times, finding an appropriate acronym is advisable, and, if you know your periodic table, you can see where this is heading. The unfortunate acronym, rendered 'CuNT' makes for some awful turns of phrase:

---

* The reason for this excessive use of obscenities is a story in itself. In the late 1960s a loosely linked group of linguists were developing a movement, 'generative semantics', with a strong anti-authoritarian streak. This generally manifested as self-deprecating humour and/or deliberate unprofessionalism, the idea being that scholars wouldn't take them seriously and they would therefore know that when their theories succeeded they would be doing so on their own merits. In that spirit, they searched for bizarre and provocative example sentences to communicate their concepts. They also wrote some of the first linguistics papers about obscenity and humour. Their movement withered, but their papers (and obscenities) remain.

- 'We electro-deposited one sample with only CuNTs inside the half depth of the nanochannels.'

- 'The CuNTs have closed caps on top.'

- 'The formation of the CuNTs depends on two factors. The first factor is gold-sputtering.'

- 'The wall thickness of the CuNTs is about 10 nm.'

The unfortunate acronym was widely reported, but the researchers continued to use it in a later paper.[89]

Intentional uses of the C-word tend to come from gender studies papers, which, given the content of the papers, is as unsurprising as it is depressing. They include '"Back to the kitchen, cunt": speaking the unspeakable about online misogyny',[90] one scholar's horrifying stocktake of just how hard it is to be a woman on the internet.*

'Motherfucker' is also mercifully underutilised in academia (though someone did write a whole book on its history).[91] It appears in the title of a book chapter about profanity in HBO's television programming,[92] and again in a book chapter about the TV show *Deadwood* (also HBO).[93] Sondre Lie from the University of Oslo wrote a thesis on the subtitling of tricky taboo words in films, giving it the title 'Translate this, motherfucker!'[94]

---

* I was absolutely mortified to find that one of the first vitriolic comments quoted by the author is attributed to another G. Wright.

SOME EXAMPLES OF WISTFUL ACRONYMS IN SCIENTIFIC PAPERS (SEXWASP)[95]

Multiple Intense Solvent Suppression Intended for Sensitive Spectroscopic Investigation of Protonated Proteins, Instantly (MISSISSIPPI)

Finnish Geriatric Intervention Study to Prevent Cognitive Impairment and Disability (FINGER)

Biodiesel Exhaust, Acute Vascular and Endothelial Responses (BEAVER)

Randomised Assessment of Treatment using Panel Assay of Cardiac markers (RATPAC)

Magnetic Resonance Imaging for Myocardial Perfusion Assessment in Coronary Artery Disease Trial (MR IMPACT)

Genetic variation and Altered Leucocyte Function (GANDALF)

Black Intelligence Test of Cultural Homogeneity (BITCH)

McGill Self-Efficacy of Learners For Inquiry Engagement (McSELFIE)

Proton Enchanced Nuclear Induction Spectroscopy (PENIS)

SearCh for humourIstic and Extravagant acroNyms and Thoroughly Inappropriate names For Important Clinical trials (SCIENTIFIC)

ACADEMIC TRANSLATOR

What academics say	What they mean
Various sources	I forgot the name and author of that one paper
We are grateful to the two anonymous peer reviewers for their constructive comments	God help them if I ever find out who they are
A promising area for an initial study	I have to do this to get funding
Widely discussed in the academic community	I accidentally ended up in the middle of a heated Twitter argument
The notes were meticulously transcribed	I was drunk and missed out at least seven pages
An extensive literature review	A quick Google search
A complex phenomenon	I don't understand
Has long evaded the understanding of scientists	I don't understand why I don't understand
Is impossible to summarise simply	I still don't understand
Approaching the traditional threshold for statistical significance	Not significant
More research is required	I need funding

OBSCURE
INTERLUDE

THE 'SCIENTIFIC' METHOD

Theory is when you know everything but nothing works.
Practice is when everything works but no one knows why.
In our lab, theory and practice are combined:
nothing works and no one knows why.

For all the pretence of objectivity and control, science can be a satisfyingly rough and ready business. There are likely thousands of scientific experiments that could be filed under 'MacGyver', but we rarely get to hear about them as the true story is lost in the transition from lab to publishable paper.[*1]

The Twitter hashtag #OverlyHonestMethods, which started in 2013, has seen thousands of contributions from academics of all disciplines sharing their slightly unscientific approaches. The tweets offer some candid insights into the day-to-day functioning of labs and offices across the world:[2]

- We used jargon instead of plain English to prove that a decade of grad school and postdoc made us smart.

---

* For example, Elyse Ireland from the University of Chester told me that for one of her team's papers on techniques for detecting human blood (for forensic applications), one of the authors personally provided the blood samples. This involved giving blood three times for replication and reproducibility purposes.

- Brains were removed and dissected in, on average, 58 seconds. We know precisely due to a long-running lab competition.

- Stimuli for this experiment were inspired by a Monty Python sketch . . . they worked so I stuck with it.

- Slices were left in a formaldehyde bath for over 48 hours, because I put them in on Friday and refuse to work weekends.

- I used that specific sequence of biotinylated DNA because I found some in the freezer.

Many media outlets, seemingly unaware that scientists (sometimes) have a sense of humour, saw the hashtag as an online confessional. But really it is about the highs and lows of academic life that scientists share: working weekends and nights because there's a deadline looming or because it is the only time that an expensive new bit of equipment is available; drinking implausible amounts of coffee; and being frustrated at the constraints imposed by funders and employers. Scientists sometimes take shortcuts, but that doesn't mean science is broken.[3]

Occasionally, published papers can be starkly honest too. Researchers on a paper about the influence of the moon's phases on sleep admit that they hadn't considered their line of argument when they collected the data, but that: 'We just thought of it after a drink in a local bar one evening at full moon, years after the study was completed.'[4] In a similarly honest fashion, a couple of French researchers studied how birds react to speeding cars: 'The study took place in western France mostly on our way home.'[5]

I find these pragmatic methodologies of convenience reassuring, almost comforting. The same cannot be said of the methods section of a 1969 study, that I can only hope would no longer pass the relevant animal ethics review procedures:

> *After unsuccessful attempts to trap the redtail monkeys at the*
> *Zika Forest with the intention of live-bleeding and release,*
> *monkeys had to be sampled by means of 12-bore shotguns.*

As disgusting, but not as depressing, Hare et al. describe in detail how they got hold of cow dung for their study:[6]

> *Fresh cow dung was obtained from free-ranging, grass-fed, and*
> *antibiotic-free Milking Shorthorn cows (Bos taurus) in the Tilden*
> *Regional Park in Berkeley, CA. Resting cows were approached*
> *with caution and startled by loud shouting, whereupon the cows*
> *rapidly stood up, defecated, and moved away from the source of*
> *the annoyance. Dung was collected in ZipLoc bags (1 gallon),*
> *snap-frozen and stored at −80°C. Dung [was] thawed at 4°C and*
> *moistened slightly before use.*

SOMETHING FISHY

In 2009, a team of neuroscientists and psychologists conducted a study wherein they showed a series of photographs depicting social situations to their subject, asked them to determine what emotion the individual in the photo was experiencing, and measured their brain activity in an MRI scanner.[7]

The sole participant in the study: 'One mature Atlantic Salmon, 18 inches long, 3.8 lbs ... not alive at time of scanning.'

This silliness started out as a standard pre-study machine test, used to calibrate the scanner.[*] Craig Bennett and his team weren't content with the low contrast scans of the oil-filled balloon commonly used for such tests. Ever the scientists, they worked their way through a menu of options. They started with a pumpkin, but were dissatisfied with its lack of compositional complexity. Then they scanned a Cornish

---

[*] The results were set aside and not revisited until much later when one of the co-authors was teaching a seminar on the proper analysis of MRI data. They needed an example of improper analysis and remembered the salmon data sitting unused on the computer.

game hen (also not alive at the time of scanning), which produced a decent image, but still wasn't as punchy as they wanted. Finally, they settled on the scan of the salmon for its rich mix of textures.

The procurement of said salmon led to arguably the most delightful declaration in the history of academia. Bennett marched into his local grocer and declared:

I need a full-length Atlantic Salmon. For science.[*]

While an ex-Cornish game hen may be useless, a salmon that has shuffled off its mortal coil and joined the choir eternal is quite the opposite: far from being bereft of life, the uncorrected scans showed activity in the salmon's brain and spinal cord. Of course, what they actually show is that improperly analysed scans could lead to the mistaken belief that dead salmon are unexpectedly pensive.[†]

---

[*]   Bennett was not reimbursed for the salmon, which was later eaten.

[†]   Here is my attempt to explain what is going on: The visual data produced in MRI scans is generally broken up into sections called 'voxels' (essentially 3D pixels). Such scans of the brain produce a *lot* of data – somewhere between 40,000 to 130,000 voxels per image. To identify the brain regions at work, two scans are compared with each other by looking at each voxel to see if it is 'activated' (i.e. if that part of the brain is firing). It is necessary to make thousands of such comparisons to generate an overall picture (and running the stats quickly becomes complex and cumbersome). This causes the so-called 'multiple comparisons problem': given the number of comparisons being made, it is inevitable that some of them will be false positives (e.g. voxels may appear activated through random noise in the equipment). During the 1990s, various methods were developed for correcting these red herrings, the most popular being to calculate the probability of a voxel being falsely activated and excluding those that are likely out of place. However, this can have the adverse effect of reducing the statistical power of the original comparisons (i.e. the false positives are removed, but true positives may also be excluded, resulting in false negatives). As such, not all neuroscientists use multiple comparisons correction when analysing their data and reporting the results. Bennett et al. argue that false negatives are the lesser of two evils and show, with salmon, that if the comparisons are left uncorrected there is a good chance you will see some brain activity wherever you look.

Thus the simple salmon shot to fame, becoming the top Google result for 'salmon study'* and the poster child for corrected scans (literally – the paper started out as a conference poster). At the time the poster was first presented in 2009, around 25–40% of published MRI studies were presenting uncorrected comparisons; by 2012, when the authors won the Ig Nobel Prize for Neuroscience, the figure was 10%.

It is not known whether the authors sought ethics approval for the study, though I understand that the salmon did not consent to its participation on account of it being dead, and a salmon.

---

* The study shares the top ten search results with just three other salmon studies: a report on the mislabelling of fish sold in restaurants (because we love salmon but would probably never order a fillet of slimehead), another on salmon aquaculture methods, and a study finding that farmed salmon get depressed.

TEACHING

'Those who can, do. Those who can't, teach.
Unless you are an academic, in which case you probably
have to teach regardless of your ability.'

Academic teaching is a strange enterprise. It requires academics, stereotypically better known for their prowess in solitary tasks, to stand in front of large groups of reluctant teenagers of varying abilities and attempt to impart the rudimentary basics of subjects they have committed their lives to becoming experts in. All of this is generally done with little to no formal training.

I am in the privileged position of only having to teach occasionally, lecturing on topics I love to receptive and enthusiastic students. As a result, my experience of teaching has been incredibly positive, if somewhat skewed. By contrast, many academics see teaching as an unfortunate but necessary obligation that detracts from their research.

I asked the academic Twittersphere to complete this sentence: 'Teaching is _____.' The following two responses best sum up the range: 'Underrated, amazing, overlooked, essential, underpaid, rewarding, tiring, and inspiring.'; and 'Dante's Seven Levels of Hell.'*[1]

---

* Note: there are nine circles of hell in Dante's *Inferno*.

Perhaps it is the sense of obligation that leads a lot of academics to resent the entire enterprise. Or possibly it is the ever-increasing teaching load, uncooperative students, and student reviews that often have little to do with the quality of the teaching.

FAIL EVERYONE

If teaching a class becomes too much, there is an out: fail the entire class. When Irwin Horwitz of Texas A&M University felt that an exceptionally awful cohort was beyond redemption, he sent the students an email:

> *Since teaching this course, I have caught and seen cheating, been told to 'chill out', 'get out of my space', called a 'fucking moron' to my face, [had] one student cheat by signing in for another, one student not showing up but claiming they did, listened to many hurtful and untrue rumors about myself and others, been caught between fights between students …*
>
> *None of you, in my opinion, given the behavior in this class, deserve to pass … It is thus for these reasons why I am officially walking away from this course. I am frankly and completely disgusted. You all lack the honor and maturity … and the competence and/or desire to do the quality work necessary to pass the course just on a grade level … I will no longer be teaching the course, and all are being awarded a failing grade.*

The same day, Horwitz sent a similar email to the senior administrators of the university telling them what he had done, that the students were no longer his problem, and predicting (correctly) that students would protest. Equally predictable was the swift response from the university – you can't just fail everyone.

In an interview, Horwitz later said that the class was his worst in 20 years of college-level teaching and he felt he had no choice after his complaints to university administrators went unanswered.[2] The move polarised academics, who either mocked him for being thin-skinned or praised him for taking a stand.

This was not the first time an instructor has taken drastic action when pushed to their limits. A philosophy professor at Syracuse University caused controversy with his policy of leaving class immediately if he spotted a student texting, while two engineering professors at Ryerson University informed students that they would be given three warnings about disruptions before the professors would walk. The university forced them to abandon the policy before they had a chance to use it.[3]

Dear Student,

I am writing to inform you that I have marked you absent for today's class, irrespective of the fact that you were physically present.

Our TA was sitting behind you during class and reported that you spent the entire class searching for pictures of 'puppy golden retrievers with party hats on' while attempting to stifle your laughter.

Important and gratifying though that activity is, I strongly advise you to do it in your own time.

See you on Tuesday.[4]

PASS EVERYONE

At the other end of the scale is the even rarer case of a university seemingly willing to pass everyone, regardless of the quality of their work. The University of North Carolina at Chapel Hill was at the centre of controversy when one of its departments was found to be providing sham 'paper classes', apparently used to keep struggling student athletes enrolled so they could continue to play for college sports teams.

The report of an independent investigation details the alleged depth and blatancy of a long-running scheme whereby students simply submitted a paper, generally of exceedingly poor quality, in exchange for the grade they needed to remain enrolled.[5] This was discussed somewhat openly amongst coaches, teachers, and other staff.[*]

The report says that department administrator, Deborah Crowder, masterminded the scheme and oversaw its running for fifteen years. She apparently graded the papers herself and awarded top marks as long as the papers met the required length. An email exchange between Crowder and Jan Boxill, who was an academic counsellor to women's basketball players at the time,[†] highlights the farcical nature of the scheme:

> **Crowder:** As long as I am here I will try to accommodate as many favors as possible. Did you say a D will do for [basket-ball player]? I'm only asking that because 1. no sources, 2. it has absolutely nothing to do with the assignments for that class and 3. it seems to me to be a recycled paper...
>
> **Boxill:** Yes, a D will be fine; that's all she needs. I didn't look at the paper but figured it was a recycled one as well, but I couldn't figure from where!

---

[*] Members of the counselling staff presented a slide to football coaches saying, 'We put them in classes that met degree requirements in which: They didn't go to class; They didn't take notes, have to stay awake; They didn't have to meet with professors; They didn't have to pay attention or necessarily engage with the material.'

[†] Boxhill is also a philosophy professor and has written books about ethics in sport.

As a result of these lax grading standards, the average GPA of the students in these classes was 3.61, compared with 1.92 in other classes. Student advisers from the athletics department maintained a list of struggling athletes and the grades they needed to stay eligible to play, and steered student athletes to the classes. More than 20% of the university's athletes from 1999 to 2011 were enrolled in these classes.

When Crowder announced she would retire in 2009, panic ensued. The associate director of the athletics advising programme wrote to a staff member that they should expect students to fail if they didn't get their papers in before she left. Following Crowder's departure, the GPA of the football team fell to its lowest level in ten years. With the eligibility of their athletes at risk, counsellors for the football team pressured then department chairman Julius Nyang'oro to continue the fake classes. He apparently acquiesced, and six more classes went ahead, one of which was taken by 13 football players.

While the scheme clearly violated basic standards of academic integrity, there is no evidence that Crowder or Nyang'oro sought to personally profit or unduly inflate the stature of their department. Indeed, the investigation suggests that their hearts were in the right place. Crowder had herself attended the University and recounted that 'she was left adrift by a faculty and staff that focused on "the best and the brightest" and failed to pay attention to students like herself who needed direction and support,' so she felt she had a duty to help others who faced similar struggles. Nyang'oro was haunted by the fates of two athletes he taught early in his career who lost eligibility and drifted – one was murdered after returning to his hometown and the other ended up in prison.

PAR FOR THE COURSE

Fronting a sham class is likely the least effort you can invest in educating future generations, but given the dubious ethical implications, a more commendable low-effort option is to teach a course on a subject that students already know inside out. In 2014, the University of

Pennsylvania's English department began offering a course entitled 'Wasting Time on the Internet', taught by eccentric academic and poet Kenneth Goldsmith.*

In the course, Goldsmith aims to use social media, cat videos, status updates, and online shopping as the inspiration for works of literature. 'Could we reconstruct our autobiography using only Facebook?' the course description asks. 'Could we write a great novella by plundering our Twitter feed? Could we reframe the internet as the greatest poem ever written?'

All the class requires is a laptop and a WiFi connection, though students also 'explore the long history of the recuperation of boredom and time-wasting through critical texts about affect theory, ASMR,† situationism and everyday life.' The course description concludes: 'Distraction, multi-tasking, and aimless drifting is mandatory.'

Taken at face value, the course may seem bizarre, but Goldsmith argues that daydreaming and distraction have long been an integral part of the creative process.

Intrepid *Slate* journalist Katy Waldman sat in on one of the seminars and reported on the following diverse activities:[6]

---

* Goldsmith wrote *Traffic*, a collection of traffic reports arranged as poetry, and read sections of it (compellingly, I might add) at a poetry event sponsored by President Obama (who can be seen laughing heartily in a video of the reading). Goldsmith's attempt to poetically remix Michael Brown's autopsy report at a conference in 2015 was less well received. Another of Goldsmith's courses, 'Uncreative Writing', promises students that they will learn to employ 'strategies of appropriation, replication, plagiarism, piracy, sampling, plundering' as writing techniques.

† A neologism meaning 'Autonomous Sensory Meridian Response': a combination of pleasurable physical and psychological affects, primarily relaxation, experienced in response to external stimuli, especially whispering or soft-spoken voices, or precise movements on a visual plane. Search the term on YouTube and you will find a great number of videos dedicated to lulling you into such a state, including titles such as 'Maria spends 20 minutes folding towels', 'Long Hair Brushing Session for Relaxation', and '~♥~ Let me take care of you ♥~'. The internet is weird.

- The '30 seconds of heaven' exercise, wherein laptops are rotated around the class, giving each student 30 seconds to open anything they like on your computer.

- Watching a YouTube video entitled 'Try Not to Laugh!!! (IMPOSSIBLE CHALLENGE!!!)', starting over every time someone chuckles.

- Applying for jobs using random CVs lifted from LinkedIn.

During the class Goldsmith reminds students to seek out the 'stuplime', i.e. where the stupid and the sublime become so intertwined that you struggle to separate the two. 'Something is so stupidly sublime or sublimely stupid that it becomes transcendent.' This stuplime state of transcendence should, he posits, allow the creative juices to flow. But this hasn't happened, yet. Goldsmith says that not one student produced anything interesting in the first few writing assignments.

Waldman concludes that the class is 'Just as provocative, infuriating and elusive as it sounds... As a concept, it shimmers with just enough promise to make the underdelivery bite.'

Clearly keen to outstuplime this American maverick, the University of Leicester used *Back to the Future Day*[*] to announce that it had established a 'Department of Transtemporal Studies'.[†][7] The course webpage promises that 'Staff in the Department have extensive experience of journeying to a wide variety of historical and future periods' and that 'Anyone studying for a degree in Transtemporal Studies can be sure of solid employment and steadily increasing wages for at least the next 50 years (apart from a brief recession in the late 2040s).

---

[*] 21 October 2015, the day that Marty McFly ends up in when he uses the time machine.

[†] The original page has now disappeared and has been replaced with a notice stating, 'After many years of studying the future, the Department of Transtemporal Studies has now closed due to unforeseen circumstances. It will reopen in 2045.' The original page can still, rather appropriately, be accessed using the Wayback Machine.

Table 5: Underwater basket-weaving and other Mickey Mouse classes[*]

Course title	University	Course description
Zombies in Popular Media	Columbia College Chicago	'This course explores the history, significance, and representation of the zombie as a figure in horror and fantasy texts. Instruction follows an intense schedule, using critical theory and source media (literature, comics, and films) to spur discussion and exploration of the figure's many incarnations.'[8]
Sport, Media and Culture (dubbed David Beckham Studies by the popular press)	Staffordshire University	'Examining the rise of football from its folk origins in the 17th century, to the power it's become and the central place it occupies in British culture, and indeed world culture, today.'[9]

---

[*] The term 'underwater basket weaving' has long been used as pejorative designation for any university course perceived as being useless or absurd, or to describe a perceived decline in academic standards more generally. In 1919 one writer lamented: 'Higher education is becoming very practical indeed. It includes everything nowadays – excepting, of course, Greek and Latin – from plumbing to basket-weaving' ('Studying National Parks', *The Watchman and Southron*, 6 August, 1919). There are many references in the 1950s (incidentally, many of these concern sham courses given to student athletes), including: 'These may include courses in life-insurance salesmanship, bee culture, square-dancing, traffic direction, first aid, or basketweaving' ('Magna cum nonsense', *New York Times*, 16 March, 1952). A 1956 edition of *American Philatelist* noted, quite seriously, in a piece on a remote Alaskan community, that: 'Underwater basket weaving is the principal industry of the employables…' The phrase later came to be used to describe courses that young men took to dodge the draft during the Vietnam War. Wanting to get in on the joke, many universities have offered one-off courses in underwater basket-weaving.

How to Watch Television	Montclair State University	'The aim is for students to critically evaluate the role and impact of television in their lives as well as in the life of the culture.'[10]
What if Harry Potter Is Real?	Appalachian State University	'This course will engage students with questions about the very nature of history ... The Harry Potter novels and films are fertile ground for exploring ... issues of race, class, gender, time, place, the uses of space and movement, the role of multiculturalism in history.'[11]
How Does it Feel to Dance?	Oberlin College	'Whether you say "I don't dance," or "I love to dance," this course is for you.'[12]
Stupidity	Occidental College	'This course examines stupidity.'*[13]

---

* If that were the entire course description, I would think it quite amusing, but the actual course description is reminiscent of the headache-inducing academic writing seen earlier: 'Stupidity is always the name of the Other, and it is the sign of the feminine. This course in Critical Psychology [is] a philosophical examination of those operations and technologies that we conduct in order to render ourselves uncomprehending. Stupidity, which has been evicted from the philosophical premises and dumbed down by psychometric psychology, has returned in the postmodern discourse against Nation, Self, and Truth and makes itself felt in political life.' Call me stupid, but I don't understand what this course is about.

Tu 9:50am to 1:00pm – Classroom: West Duke 105 – Office: 255
Sociology/Psychology
Soc 710
Social Theory Through Complaining
Kieran Healy, Duke University

This course is an intensive introduction to some main themes in social theory. It is required of first year PhD students in the sociology department. Each week we will focus on something grad students complain about when they are forced to take theory. You are required to attend under protest, write a paper that's a total waste of your time, and complain constantly.

Passive-aggressive silence will not be sufficient for credit.

Course Schedule

Week 1	*Introduction: This has Nothing to do with my Research Interests*
Week 2	*This is all just Obfuscatory Bullshit and Empty Jargon*
Week 3	*It's Not Like We Can Even Predict Anything*
Week 4	*Isn't it more Complicated than that?*
Week 5	*Aren't these things Mutually Constitutive?*
Week 6	*But what about Power?*
Week 7	*We could easily Fix this Mess with some Basic Math*
Week 8	*This Field is Sexist and Racist to its Rotten Core*
Week 9	*What is Theory without Praxis?*
Week 10	*THANKSGIVING BREAK. If You Can Call it a Break.*
Week 11	*Look, if Everything is Socially Constructed, then Nothing is*
Week 12	*Can you Believe we didn't Read any _____?*
Week 13	*Conclusion: This Whole Project was an Exercise in Symbolic Violence*

READ THE SYLLABUS

As the number of students in a class increases, the probability that someone will ask a question that is already answered in the syllabus approaches one. Exasperated, a few cheeky teachers have taken to testing whether the class is reading the syllabus by inserting unusual requests:[14]

- Joseph Howley, an assistant professor of classics at Columbia University, asked students to email him a picture of the character Alf from the popular eighties sitcom ALF (with the subject line, 'It's Alf!'). He said that the Easter egg 'yielded quantitatively dismal results', but had nonetheless resulted in some amusing emails.*

- Damian Fleming, an associate professor of English and linguistics at Indiana University-Purdue University, asked his students to send him a picture of a 'cool medieval tattoo'. Around half of his students humoured him with a response.

- Adrienne Evans Fernandez, an adjunct professor of biology at Ivy Tech Community College, in Bloomington, Indiana, asked her students to send her a dinosaur picture. About 25% of students did.

MAKING THE GRADE

Perhaps the only aspect of academic life more maligned than teaching is grading. While grading is unlikely to become exciting anytime soon, there are a couple ways academics have tried to make it interesting.

Every year since 2008, Professor Dylan Selterman of the University of Maryland has presented his class with a prisoner's dilemma:[15]

---

* One student noted the apparent contradiction between the request and the edict that email 'should be approached as a professional communication'.

> *You can each earn some extra credit on your term paper. You*
> *get to choose whether you want 2 points added to your grade,*
> *or 6 points. But there's a catch: if more than 10% of the class*
> *selects 6 points, then no one gets any points.*

Meanwhile, a screenshot of the following grading policy has been doing the rounds on social media:

> *Some of you think that attendance is not necessary to pass*
> *a college course. I don't know who you are. I don't know*
> *what you want. If you're looking for an easy A, I can tell*
> *you I don't have your easy A but what I do have are a very*
> *special set of skills. Skills which I have acquired over a*
> *very long career. Skills that make me a nightmare for people*
> *like you. If you attend classes regularly, that will be the end*
> *of it. I will not look for you, I will not pursue you. But if*
> *you don't, I will look for you, I will find you, and I will*
> *grade you.*

Academics' disdain for grading is equalled by student superstitions surrounding exams. At Royal Holloway, University of London, a painting hanging in the exam hall is shrouded in superstition. The painting depicts the mysterious demise of Sir John Franklin's fabled 1845 Arctic expedition, showing two polar bears devouring the remains of a ship and its occupants. Ever since the first exams in the 1920s the painting has been associated with failure. 'If you sit directly in front of it in an exam, you will fail – unless it's covered up,' says Laura MacCulloch, the college's curator.[16]

In the 1970s a student refused to be seated near it and a massive Union Jack was found to cover it. The flag has adorned the painting every exam period since. The legend has morphed over the years, with a recent version being that a student had stared directly into one of the polar bears' eyes, fallen into a trance, gone mad and killed herself, though not before etching 'THE POLAR BEARS MADE ME DO IT' onto her paper.*

Superstitions surrounding exams elsewhere are less macabre. In the town of Göttingen in Germany, recently minted doctoral graduates rush off to kiss a statue of Lizzy, aka 'Goose Girl', at a fountain in front of the medieval town hall, while in Wisconsin students have been placing plastic pink flamingos on the main lawn at graduation since 1978.

RATE MY PROFESSORS

Rate My Professors (RMP) has exploded in popularity since first being launched in 1999. For its target audience it is a godsend, with students logging on to figure out where the easy grades are, or, less cynically, where they might get a great learning experience.†[17] For academics it can be a mixed bag. Often it is more berate than rate, and RMP has confronted many an academic with the uncomfortable truth that they aren't as popular with their students as they thought.

Reviews calling professors 'useless' or a 'general moron' are common, and relatively polite compared with:

- '…horrific teacher. No one shows up to class because it's so miserably boring. When I actually do go to class, halfway through i begin to hate God for giving me the legs that brought me there. You could walk into this class rolling on E, and by the time the second slide comes up, you'd be sober.'

- 'Once or twice, his theory talk was interesting, but other than that the only thing that keeps the blood in my brain flowing is wondering what the hell is up with the fanny pack.'

---

* Well, that escalated quickly.

† Studies confirm that this cynicism is warranted: there is a strong correlation between students' rating of easiness and quality on the website, i.e. students perceive easy lectures to be of better quality than hard ones.

- 'Whatever you do ... AGREE with her on ALL issues, praise her and tell her she is the greatest, fall down to your knees and worship her, then maybe, just maybe you might make a B.'

- 'Take him if you need the class. But come prepared with an energy drink and a coloring book because that is the only way you will last.'

- 'If I had a choice between taking another one [of his] classes and being saturated with brown gravy and locked in a room with a wolverine that is high on PCP, then I honestly believe that I would choose the latter.'*

Not only are these all real reviews, I could have filled an entire book with them.

While some in the academic community are understandably critical of the site and dismissive of such venomous evaluations, professors at York University's Osgoode Hall Law School in Toronto and Simon Fraser University, British Columbia had a better idea.[18] No doubt inspired by a popular segment on the *Jimmy Kimmel Show* where celebrities read out nasty tweets accompanied by REM's 'Everybody Hurts', they posted videos of staff reading their negative RMP reviews.

'Hes hot in during the lecture, but after lecture hes super cold,' reads Peter Tingling, associate professor of management information systems in one video. 'Before I attended his class, I thought he was a women prof,' says Enda Brophy, a male assistant professor of communications.

The deadpan deliveries are the best. 'I found this course to be tediously boring, and Steve was useless, although he is a very nice guy,' reads

---

* Phencyclidine (PCP), also known as angel dust, is an anaesthetic, brought to market in the 1950s, but banned in 1965 due to the high prevalence of dissociative hallucinogenic side effects. It continues to be distributed illegally as a recreational drug. PCP can numb the mind, cause aggressive behaviour, and induce feelings of strength, power, and invulnerability. You do not want to be locked in a room with a wolverine that is high on PCP.

Stephen Collis, professor of English. 'Consolation prize,' he says, smiling and giving a cheeky nod to the camera. 'Awfully boring class if you're not interested in environmental engineering,' reads Kristen Jellison, associate professor of environmental engineering who was teaching 'Introduction to Environmental Engineering'.

Todd Watkins, a professor of economics, reads a review referring to the university's Integrated Product Development (IPD) programme: 'This says I'm useless to the IPD programme and a general moron... Hell, I started the dang IPD programme!' The reviewer then complains that Watkins rambles too much, to which he responds by rambling quite deliberately to the camera.

Dannagal Young, associate professor of communication at the University of Delaware, appeared in a RMP video produced by students at her university, though in her case the shoe is on the other foot: her reviews are uniformly positive, and in the video she pretends to mock her students for liking her course. Young's research is on the uses of political humour and satire, and she reckons that the key to making such videos funny is to find a suitably offensive comment and 'own those sentiments proudly... Once empathy is activated, it undercuts the joke.'

Benjamin Schmidt, an assistant professor of history at Northeastern University, created a tool to identify the frequency of word usage in RMP reviews by discipline, using a database of words drawn from 14 million reviews. The tool was intended to highlight differences in how students address male and female faculty, and even a brief dabble can be quite disheartening.[♀19]

Several positive words, in terms of academic reputation, appear far more frequently in reviews of male professors than of female professors: 'Smart', 'intellect', and 'genius' all appear with greater frequency in reviews of male professors in all 25 disciplines for which data is available.[*]

---

[*]   I know 'data' is plural, but 'data are' just doesn't sound right to me. English is a flexible language, so it may be time to accept that 'data' is a welcome exception to the strict rules (but if you try to take away my Oxford commas, you will have to pry them from my cold, dead, and lifeless hands).

Words more commonly found in reviews of female faculty tend to fit certain stereotypes, both positive and negative, such as 'bossy' and 'nurturing'. Fashion-related words are also common, and female profs are also more likely to be called 'demanding', except in a few disciplines.

Not all words are so strongly gendered though, and there are some less predictable gender differences in word choice: female professors are more likely to be called 'mad' and 'crazy', while male professors are simultaneously seen as more 'funny' and 'boring'. The descriptors 'dumb' and 'stupid' remain satisfyingly gender neutral.

Gender imbalances aside, a lot can be learned about the academy from typing in random phrases. The physics faculty is top of the class for hairiness due to an unexplained preponderance of hairy females, while the hairiest men are overwhelmingly in education and philosophy. A search for 'bad teeth' reveals a high prevalence of odontophobia among male anthropologists and female historians. 'Irritating' professors are to be found in anthropology, fine arts, and communication, while 'awesome' professors teach criminal justice and psychology.

Even the most unlikely words and phrases have been used in a review somewhere. The terms 'tea bag', 'sand castle', and 'baby food' all make an appearance for example.*

LET THE GAMES BEGIN

As if the barbs of disgruntled students on RMP weren't enough, a Republican Iowa State Senator tabled an ill-considered bill targeting professor performance that the President of the American Association of

---

* These terms appear in the following reviews: 'Biggest tea bag ever. She never helped anyone in office hours and didnt teach anything that was covered on her exams. SWITCH TO ANYONE ELSE. if you take her, your done.'; 'He wears the same shirt for weeks and likes to play with the chalk and rub it in his hair and afterwards he likes to drain his tea bag with chalky hands.'; 'His lectures are death but make sure to listen and read slides. Midterm is hard, final is baby food.'; 'In a word, BORING. His is the kind of creative genuius it takes to build a sand castle.'

University Professors called, 'The most outrageous proposal I have heard from a legislator anywhere.'[20]

The bill would have required professors at public institutions to be rated by student evaluations, and goes on to say that if the professor fails to attain a minimum threshold of performance based on the student evaluations, the institution shall terminate the professor's employment regardless of tenure status or contract.

One bizarre provision is more reminiscent of *The Hunger Games* than higher education: the bill would have instituted a system of public voting to decide whether to terminate the employment of professors that met the minimum standards, but were in the bottom five performers. According to the proposal, their names would be published on the institution's website and students would vote. The professor with fewest votes would then have their contract terminated, regardless of tenure status.

The bill died a swift death in committee, but nonetheless exemplifies the growing student-as-customer mindset that has many academics worried.

OBSCURE INTERLUDE

FOOD, GLORIOUS FOOD

Crisper: If you've ever found yourself peckish with nothing to hand but half a bag of stale crisps, there is a simple solution for turning them into an appealing snack: play crisp noises while you eat.[1] This tricks your brain into believing that they are fresh.

Bowled over: In one study on the link between appetite and portion size, participants slurped soup from bowls that quietly refilled themselves over a twenty-minute period.[2] Researchers wanted to measure whether participants ate more from the refilling bowls (of course they did). My university is yet to respond to my urgent request to equip our offices with a ramen noodle delivery system based on this model.

Use your noodle: The University of Rochester offered one imaginative student a place at the university after he wrote an impressive admissions essay on his love for ramen noodles.[3]

The cheek of it: The medical literature is replete with stomach-churning accounts of food-related mishaps. In one case, a Korean woman complained of a prickling sensation in her mouth after eating a portion of parboiled squid.[4] The doctor found 'twelve small, white spindle-shaped, bug-like organisms' attached to the inside of her cheeks and tongue. These turned out to be the 'parasite-like sperm bags' that the squid would have otherwise deployed for mating purposes.

Don't play with your food: One case report documents a man with lipoid pneumonia, caused by injecting olive oil into places he shouldn't have.[*][5] Another demonstrates that even a salami can be dangerous in the wrong hands (though 'Rectal Salami' is a truly incredible paper title).[6] It is, however, occasionally acceptable to stick food where the sun doesn't shine: one report recounts the fashioning of a 'nasal tampon' from cured pork to stem a nosebleed.[7]

Fish face: There is a rich literature on the swallowing of whole live fish, with at least four reports of this unfortunate error. One such report, entitled 'Return of the Killer Fish', documents the case of a 45-year-old man who, while drinking on a fishing trip with friends, attempted to swallow a whole live fish and died from asphyxiation.[8]

Piece of cake: In his book *Admissible Sets and Structures* Jon Barwise writes: 'Section 6 should be supplemented with coffee (not decaffeinated) and a light refreshment. We suggest Heatherton Rock "Cakes".' He provides a recipe, reassuring readers that they 'taste better than they sound'.[†]

---

[*] On reflection, I am not sure that you should be injecting olive oil into any of your body parts.

[†] Recipe: Combine 2 cups of self-rising flour with 1 teaspoon of allspice and a pinch of salt. Use a pastry blender or two cold knives to cut in 6 tablespoons of butter. Add ⅓ cup each of sugar and raisins (or other urelements). Combine this with 1 egg and enough milk to make a stiff batter (3 or 4 tablespoons of milk). Divide this into 12 heaps, sprinkle with sugar, and bake at 205°C for 10–15 minutes.

A lovely cup of tea

As a British tea-drinker working in France, I have struggled with my choice of hot beverage. The social pressure to drink coffee here is as overpowering as the coffee itself, and there is no communal milk in the office (there is, however, a cupboard containing a seemingly endless supply of olive oil, salt and balsamic vinegar).

Nonetheless, I patriotically persist, following the sage advice of the UK Ministry of Munitions (1916): 'An opportunity for tea is regarded as beneficial both to health and output.'

Many have weighed in on how to make the perfect cup of tea. The Royal Society of Chemistry has produced guidelines that recommend loose-leaf Assam, steeped in fresh-boiled* filtered water in a pre-warmed pot, complemented with milk and white sugar.[9] Neuroscientist Dean Burnett (author of the fantastic book *The Idiot Brain*)[10] concludes that the mere premise of the age-old question is itself so subjective that it can never be definitively answered.[11]

An emerging field of scientific inquiry is now considering post-brew best practice. In one paper, scientists have modelled the 'teapot effect' (the pesky dribble down the underside of the spout),[12] and in another (ironically written by four Frenchmen) have identified a few factors that affect dribbling.†[13] These include the curvature of teapot lip, the

---

* Reboiling reduces the oxygen content of the water, affecting the flavour of the tea.

† I don't mean that they wrote it with the intention of being ironic in the classical Ancient Greek comedic sense (traditional use of the term is rooted in the Greek comic character Eiron, a smart underdog who repeatedly triumphs over the boastful character Alazon), but rather that Frenchmen writing in such detail about a quintessentially English occupation is ironic. At this point, about 50% of readers are mentally screaming at the page: 'That isn't real irony! It is just an amusing contradiction between your expectations and the reality!' In fact, 'irony' has been used to describe situations that are incongruous with expectation since at least 1640 (sometimes distinguished as 'situational irony', 'irony of fate', 'irony of events' or 'irony of circumstance'). Alanis Morissette fans unite!

flow rate, and the 'wettability' of the teapot material. The main culprit, the 'hydro-capillary' effect, can easily be overcome by either thinning the spout or by applying super-hydrophobic materials to the lip.

A fraught walk back to the desk follows the making of any hot drink, with its inevitable hand-scalding and mess-making. The authors of 'Walking with Coffee: Why Does It Spill?' are sympathetic.[14] They conducted an experimental study on beverage spillage, controlling for various walking speeds and initial liquid levels, and figuring out how to stay within the 'critical spill radius' (i.e. the edge of the mug).

Some Australian researchers investigated the rate at which teaspoons disappeared from their staff kitchen by meticulously tracking 70 teaspoons for five months.[15] Teaspoon half-life was 81 days, with a staggering overall attrition rate of 80%. The researchers were stumped as to why this occurs, offering 'escape to a spoonoid planet' as one possible explanation.

Academics have even overthought the simple biscuit. 'Washburn's Equation' has been used to describe how liquid moves through the biscuit, while a team of mechanical engineers led by Len Fisher used a gold-plated digestive to figure out how best to dunk.[*][16] A full cup and an angled entry are essential, but the secret is to flip the biscuit post-dunk so that the drier side supports the weaker side as you move from mug to mouth.

Cheers!

---

[*] The research was funded by McVitie's.

IMPACT & OUTREACH

Impact in academia is like sex: everyone is talking about it, but few are having it. Or at least not as regularly and as intensely as they'd like. We all want more of it, and many of us are obsessively measuring and analysing it.[*]

An oft-repeated pearl of wisdom is that you can't manage what you can't measure,[†] and measuring impact is no mean feat. The traditional measure is citations, which is in theory as simple as counting the number of times a given paper has been cited by other papers. But it's harder than it seems. There is an entire field dedicated to measurements like this, bibliometrics, and researchers have written countless papers trying to figure out how to efficiently and accurately count citations.

In spite of this fixation on citations, there appears to be some truth in the adage that around half of all academic papers are read by just a handful of people.[‡1] For example, one study concluded that if you exclude

---

[*]   Might have overstretched the metaphor there.

[†]   In fact, this is a common misquote of a passage from W. Edwards Deming's 1993 book *The New Economics*. What Deming actually said is: 'It is wrong to suppose that if you can't measure it, you can't manage it – a costly myth'.

[‡]   But because citation analysis is complex and because any statistical analysis always depends to some extent on how you cut the data, we don't really know the exact figures.

self-citations (i.e. academics citing their own papers), approximately 80% of journal articles in the humanities don't get cited within the first five years.[2] (The figure for the natural sciences is considerably better at 27%).

These 'simple' measures of impact are not nearly nuanced enough: the total number of citations amassed by an academic can easily be increased (by self-citation, participation in a single highly-cited study, or by churning out loads of papers that each get a few citations); while referring only to the total number of papers fails to account for the quality of the work. As a result, a raft of alternatives has been proposed.

The h-index, which was set out in 2005 and is now one of the core measures of citations, attempts to measure both productivity and citation impact. It is based on the set of the scientist's most cited papers and the number of citations they have received, such that an h-index of twelve means that twelve of the academic's papers have been cited at least twelve times.[*]

There are around one thousand scholars that boast an h-index of over 100 (i.e. they have published at least 100 papers that each have at least 100 citations each).[3] American neuroscientist Graham Colditz, known for his research on obesity, currently has a world-beating h-index of 264.

Needless to say, the h-index, and all of the other proposed alternative metrics for impact, suffer from their own problems, and scholars are increasingly wondering whether such measures are not virtually meaningless in the real world. In Einstein's words: 'Not everything that counts can be counted, and not everything that can be counted counts.'[†]

---

[*] Hirsch suggests that in physics an h-index of around 12 may be typical for getting tenure as an associate professor at a major research university

[†] Despite frequent reproduction, it appears that Einstein never actually said this. The phrase instead appears to come from William Bruce Cameron's 1963 book *Informal Sociology: A Casual Introduction to Sociological Thinking*, wherein he states: 'It would be nice if all of the data which sociologists require could be enumerated because then we could run them through IBM machines and draw charts as the economists do. However, not everything that can be counted counts, and not everything that counts can be counted.'

Many national funding bodies and review processes are now starting to ask for evidence of 'societal impact' as a complement to the traditional metrics. While encouraging scholars to step outside the ivory tower and bring research to the real world might not be such a bad idea (and of course, many are already making considerable effort to do so), some dread the thought of such an outward-facing exercise. Even the term 'impact' is now often jokingly analogised with that of a car crash.

Yet there are countless ways to make an impact. Browsing through the case studies submitted to the UK's Research Excellence Framework process (the REF),* the amorphous nature of 'impact' in the modern academy is evident. In an excellent example of science and humour working together, Oliver Double at the University of Kent wrote and performed a stand-up comedy performance entitled *Saint Pancreas* to teach people about type I diabetes.[4] Elsewhere, a team of researchers at Coventry University set out to improve land management in Africa and ended up reframing an invasive tree species as a useful commodity. The government of Kenya subsequently built a green power station run on charcoal from the trees, while the Mesquite Company in Texas is now making $150,000 a year from selling the stuff for use in barbecues.[5]

---

* Despite including some elements of societal impact and outreach, the REF remains a heavily citation-focused process. My good friend Dr David Hayes described it to me as follows: 'It's a rather large-scale quality-measuring exercise for the research outputs of British academics (so as you can imagine most everyone hates it because you can't measure quality, etc. etc. etc.) which, in practical terms, dictates things like promotions, availability of academic jobs, and the amount of money universities have to throw around. Under the last REF in 2014, Universities had to nominate a selection of research staff who would each submit four pieces of research (but that was used by many institutions to cherry-pick its best and brightest and thereby massage the figures). The REF sets out criteria for grading the papers: 4* = "world-leading"; 3* = "internationally excellent"; 2* = "nationally excellent" and 1* = I forget the euphemism, but shit. Accepted wisdom is that the best pieces for REF submission will have to fall into the 3*–4* range to be competitive. And then we get into league tables and all that poisonous bollocks.'

ERDŐS

The Erdős number pays homage to the improbably prolific Hungarian mathematician Paul Erdős, whom *Time* called 'The Oddball's Oddball'.[6] Erdő"s spent his life as a vagabond, constantly travelling between scientific conferences, universities and the homes of colleagues around the world. He could fit most of his few possessions into a single suitcase, and earned enough as a guest lecturer and from various awards and prizes to fund his travels and basic needs. He donated the rest to worthy causes and people in need.

Erdős would typically show up unannounced at a colleague's doorstep, announce 'My brain is open', and stick around for long enough to collaborate on a couple of papers before moving on a few days later ('another roof, another proof').

Erdős drank copious quantities of coffee. He also took amphetamines, which he felt were an essential part of his productivity. A friend once bet Erdős $500 that he could not abstain from amphetamines for a month. Erdős easily won the bet, but complained that mathematics had been set back by a month during his abstinence: 'Before, when I looked at a piece of blank paper my mind was filled with ideas. Now all I see is a blank piece of paper.'* He promptly resumed his amphetamine use.

Erdős's publication list stretches to a face-melting 1,525 articles, and he collaborated directly with 511 people. It is from this incredible productivity and collaboration that we get the Erdős number, which describes a person's degree of separation from Erdős himself, based on their collaboration with him, or with another who has their own Erdős number. Erdős has number 0, immediate collaborators have an Erdős number of 1, and their collaborators have an Erdős number of 2, and so on. The number was first defined in 1969 by analyst Casper Goffman (Erdős = 2).[7] About 268,000 people have a finite Erdős number and, due to interdisciplinary collaborations, numerous academics in non-mathematical fields also have Erdős numbers.[8]

---

*   Join the club.

Unusual characters who might be said to have an Erdős number include:

- Matt Damon. *Good Will Hunting* was conceived and scripted in part by Matt Damon. Mathematician Dan Kleitman (Erdős = 2) was a consultant on the film, which, if you stretch the concept a bit, gives Damon an Erdős number of 3.[9]

- Baseball Hall of Famer Hank Aaron.* Carl Pomerance, a professor at Dartmouth College and one of Erdős's collaborators, reports that a baseball was autographed by Erdős' and Aaron during a ceremony to award them both honorary degrees at Emory University in 1995.[10]

- F.D.C. Willard (a Siamese cat that ended up in an author list – see page 195).[11] According to a thread on Reddit, that most reliable of sources, Willard has an Erdős number of 7.

- A horse. Jerry Grossman of Oakland University, founder of the Erdős Number Project, contributed an article to a magazine jointly with Smarty, his wife's horse. As Grossman has an Erdős number of 2, Smarty has an Erdős number of 3.[12]

An extension of the Erdős number, and a deeper dive into the small-world phenomenon that feeds it, is the Erdős–Bacon number. This is the sum of one's Erdős number and their Bacon number, i.e. the number of links, through roles in films, by which a person is separated from actor Kevin Bacon. For example, Stephen Hawking has an Erdős–Bacon number of 7: his Bacon number of 3 (via his appearance alongside Patrick Stewart in an episode of *Star Trek: The Next Generation*) is lower than his Erdős number of 4.

---

* Aaron was the baseball player who broke Babe Ruth's home run record.

K-INDEX

Neil Hall, a biologist at Liverpool University, proposed a tongue-in-cheek alternative to the h-index: the Kardashian Index. Hall was concerned that social media has made it possible to be 'renowned for being renowned', rather than for making any substantive scholarly contribution. In response, he developed a metric to 'clearly indicate if a scientist has an overblown public profile so that we can adjust our expectations of them accordingly.'[13]

The K-Index compares the number of followers an academic has on Twitter with the number of citations to their peer-reviewed work. Those with a high ratio of followers to citations (a K-index > 5), are labelled 'Kardashians'. A high K-index is, Hall says, a warning to the academic community that a researcher may have 'built their public profile on shaky foundations', while a low K-index suggests that a scientist is being undervalued.

Hall's paper is funny and worth a read. However, as a big believer in the value of social media, especially for early career researchers, I can't help but feel that Hall might be 'punching down' at those of us with less established careers than his. Either that, or Hall simply shares a misapprehension of social media common among established scholars.*

Neuroscientist Micah Allen writes:[14]

> *We (the Kardashians) are democratizing science. We are filtering the literally unending deluge of papers to try and find the most outrageous, the most interesting, and the most forgotten, so that they can see the light of day beyond wherever they were published and forgotten . . . Wear your Kardashian index with pride . . .*

---

* I wouldn't have written this book if Twitter wasn't great for fooling around and procrastinating. But I've also used it to build a network of academics in my field, get access to paywalled papers, seek support and mentorship, find co-authors, and get feedback on my work.

This is far from the only use for social media, but as someone that spends an inordinate amount of time seeking out outrageous, interesting and forgotten papers, I strongly sympathise with this sentiment.

The last word comes from another Hall, Nathan Hall of McGill University and of *Shit Academics Say* fame (see page 162). He neatly sums up the tension between the social media savvy scholars and the old guard:

> *Perhaps the most interesting thing about academics and social media is that the most traditionally influential feel above it, leaving almost completely unattended a massive lane of influence for those not asleep at the wheel.*

ALTERNATE SCIENCE METRICS

Merely hours after Hall's paper on the K-Index was published, a hashtag was born to parody it.[15] Under the banner of #AlternateScienceMetrics, the academic Twittersphere created hundreds of joke impact measures that saw a range of fictional characters, books, and films turned into elaborate metaphors for academic publishing.

The Kanye Index =

self-citations ÷ total citations[16]

Just as Kanye thinks he's the greatest rock star alive,* plenty of academics seem to love themselves a touch too much. The Kanye Index measures the level of self-citation in an author's work.

---

* He's not.

The Priorities Index =

dead house plants (HP) ÷ (total HP + total publications)[*17]

Academics are often working so hard that they neglect everything else, from house plants to relationships. Calculating your Priorities Index might just help you get some perspective.

The Minion Index =

papers you do all the work for, but end up as n^{th} author (where n is > 1)[18]

The Minion Index will likely appeal to PhD students and postdocs, who are frequently required to slog away on papers only to place 2nd or 3rd (or 9th) place on the author list.

The Bechdel Index =

papers with >2 female co-authors[19]

The Bechdel Test was originally proposed, albeit as a bit of sarcasm in a cartoon strip, to highlight the lack of films that feature women as people.[†] The test could feasibly be used to highlight academia's yawning gender gap.

The Adam Sandler Index =

identical papers published with different titles[20]

Another classic technique in academia: repackaging something you already published as something all new and shiny for submission to another journal (much like the unending stream of tediously unfunny Adam Sandler films).

---

* I particularly like this one as I have a terrible record with houseplants. I was once gifted a houseplant called 'Thrives on Neglect', which I neglected to death in a few short weeks.

† Alison Bechdel's comic strip *Dykes to Watch Out For* (1985). The Bechdel Test as originally conceived simply requires that a work of fiction feature at least two women who talk to each other about something other than a man. Incredibly, only about half of all films pass the test.

The Dawkins Index:

times quoted in internet arguments ÷ total publications[21]

The Dawkins Index identifies those whose quotes and witticisms have begun to overshadow their original academic work.*

SELF-CITATION

If impact is like sex, then self-citation is . . . an inevitable and healthy part of academic writing, in moderation. But excessive self-citation, while unlikely to cause blindness, can make you look crass and unprofessional.

Cyril Labbé, identified earlier as the cataloguer of published SCIgen papers, has also shown how easy it is to artificially inflate your academic ego using the internet. He invented an academic persona 'Ike Antkare' and generated a hundred papers, all citing each other. In this way, Antkare managed to garner a highly impressive h-index of 94 (lower than Freud, but higher than Einstein).[22]

A small number of academics, for whom collecting citations and massaging their ego via impact has become something of an obsession, have been using similar techniques to ensure that their numbers are ever-increasing.

I found the Google Scholar page of one young and celebrated professor bursting with 6,000 citations, almost all of them self-citations. The most incredible examples are the contributions of the professor's team to conferences. In one year alone the research group published six papers at a single conference, with the number of self-citations in each ranging from 25–40, totalling 150 citations out of one conference. Not bad for a

---

* Though these days it is Richard Dawkins's own social media missteps that have begun to overshadow his original work, and the (in)famous evolutionary biologist has experienced something of a fall from grace due to his propensity to send cringeworthy tweets to his 2 million followers. Dawkins has unhelpfully weighed in on the controversy surrounding Ahmed Mohamed (the young Muslim student whose home-made clock was mistaken for a bomb), suggested that some rapes are not as bad as others, and accidentally (ironically? surreptitiously?) posted a QR code with a link to a racist website in it.

day's work. In one of these papers the authors self-cite over 20 papers in the first footnote.

Another example of impact inflation was brought to my attention by Jason McDermott (the awesome artist behind the cartoons in this book). He was searching gene names in a database and started to notice a pattern: a string of publications characterising different genes looked suspiciously similar. Their titles were essentially the same, just substituting the relevant gene name each time, all had at least two core authors, and most were published in a handful of journals with relatively low impact factors. Many of the papers were rehashed digests of information obtained from existing databases, combined with some basic information about potential applications in cancer or biomedicine. The main author of these papers has published 99 in the *International Journal of Oncology*, with the self-citations generating an h-index of 48. There are also 99 papers in the *International Journal of Molecular Medicine*, with an only slightly less impressive h-index of 37. A combined search for the three core authors retrieved 216 publications with a combined h-index of 56, a number that would make any academic proud.

While excessive self-citation is routinely denounced, female academics may be failing to win chairs because they do not cite themselves enough.* Barbara Walter, of the University of California, San Diego, argues that female scholars do not cite their own previous work as much as male colleagues. This diminishes their perceived importance and prejudices them when it comes to decisions on top-level positions. To test her hypothesis, Walter and her team reviewed around 3,000 articles in the top 12 peer-reviewed political science journals. While any given publication was cited an average of 25 times, those with an all-male author list garnered an average of five more citations than those with an all-female list.♀ Walter has not yet figured out why this is, though anecdotal evidence suggests that female academics tend to look

---

* Chairs in this context refers to the highly sought after academic position – there is no academic contest to win physical chairs (yet).

unfavourably on self-promotion (and studies regarding self-promotion more generally seem to support this).[*]

IN A JIF

> *'Like nuclear energy, the impact factor is a mixed blessing.'*
> Eugene Garfield

Journals like to show they have an impact too, and for this we have the Journal Impact Factor (JIF) which counts the average number of citations made to papers published by a given journal.[23] Eugene Garfield, who is regarded as the father of bibliometrics, first mentioned the idea of a JIF in *Science* in 1955, and originally calculated them manually by noting all citations made that year in a (presumably huge) notebook.[†]

Thomson Reuters subsequently managed to get a monopoly on JIFs, and once a year the world of academic publishing waits with baited breath to see who's who. The rest of us look on and try to pretend that we don't care[‡] and that impact factors don't mean anything anyway.[§]

On calculating the impact factor for a given journal, *C&EN Onion* jokes that:[24]

---

[*]   It is always possible to find exceptions that more or less prove the rule. One high-profile case of a female scientist firmly shuns the trend: inflated stats were the shaky foundation for her career, which crumbled when she later committed scientific misconduct and embezzlement. Over half of her 4,000-plus citations were self-citations.

[†]   In a similar fashion, early bibliometric scholar Derek de Solla Price manually noted all the citations from the *Philosophical Transactions of the Royal Society* to track the exponential growth in scientific publishing. He published the seminal book *Little Science, Big Science* (1963) based on this work. As with many landmark works, this came about by accident – when he arrived in Singapore to do a postdoc, the library was not yet functional and a full set of *Transactions* was one of the few complete resources available.

[‡]   Even though we do a little bit.

[§]   Even though they do a little bit.

> *The current standard impact factor model used by scientists*
> *relies on the International Impact Factor Prototype (IIFP), a*
> *physical copy of the latest issue of the New England Journal*
> *of Medicine, stored in a climate-controlled vault under*
> *armed guard – defined as precisely 55.87(3) IF.*

Just as authors are occasionally overzealous in citing their own work, some journals have engaged in masturbatory self-referencing to bulk up their numbers. In his 1999 essay 'Scientific Communication – A Vanity Fair' Georg Franck warned that obsessive citation-counting could result in editors pushing authors to manipulate their counts by requiring citations to the journal as a prerequisite publication. Years later this fear is becoming a reality, at least in certain corners of academic publishing.

In one survey of almost 7,000 researchers, one in five said that editors had asked them to increase citations to their journal, without pointing to any specific or relevant papers, or suggesting that the manuscript was lacking.[25]

This is bad form. I've even seen an 'instructions for authors' page that told authors to cite articles from the journal, subscribe to it, and encourage their colleagues and institutions to do the same. Another journal published an annual review article citing every single paper published in the preceding 12 months, thus ensuring that each paper had at least one additional citation for that year.

While shifty strategies may work for a while, Thompson de-lists journals with unhealthy self-citation rates. For example, the *World Journal of Gastroenterology* received its first impact factor in 2000, pegged at a modest 0.993. A year later it was up to 1.445 and by 2003 it was at 3.318. The journal's success was being fuelled by self-citations, which accounted for over 90% of its total, and it was subsequently de-listed. It was re-listed in 2008, this time with a more muted impact factor of 2.081 (comprising just 8% self-citations).[26]

Over 50 journals were removed from the list in 2011 for extreme self-citation, including *Cereal Research Communications*, which had a 96% self-citation rate. It's enough to make you choke on your Cheerios.

OBSCURE INTERLUDE

SPOOKY SCIENCE

Crime writers often refer to the 'smell of death' lingering in the air after a grisly murder scene is encountered. Science tells us that decay starts four minutes after death,[1] and produces a smell comprising a complex bouquet of more than 800 'cadaveric volatile compounds'.[2] In a *PLOS One* study, a team of researchers 'sniffed' a decaying pig* using 'comprehensive two-dimensional gas chromatography-time-of-flight mass spectrometry' (which I bet sounds much cooler than it really is). Another study investigating this topic was published in the journal *Analytical and Bioanalytical Chemistry,*† but failed to detect two compounds – cadaverine and putrescine – as these are only found in decaying human cadavers and not pigs.[3]

In 2012, many doomsday evangelists predicted the end of the world, coincident with the end of the Mayan calendar.[4] Paul Wheatley-Price et al. wrote a paper considering how research might be affected by our then-imminent extinction.[5] While they argue that clinical trials would become useless in the absence of human subjects, their computer modelling shows that population actually begins to increase in the immediate aftermath of the apocalypse, even when

---

* This paper provides another example where the subject matter provided the authors with an opportunity to include a horrific graphic.

† The journal rather satisfyingly abbreviates to *Anal Bioanal Chem* when using some style guides.

controlling for known sources of bias.[*] The only plausible explanation, they conclude, is a post-apocalyptic zombie repopulation.

While the world did not end in 2012, zombies, and other mythical or undead beings, remain a concern. A paper in *Skeptical Inquirer* aims to explain away zombies, ghosts, and vampires with the power of maths and physics.[6] The authors argue, for example, that cold chills caused by ghosts are simply due to poor insulation, and note the amusing paradox that ghosts are often portrayed as walking, despite having no physical body.[†]

Vampires can be proven not to exist with some simple mathematical modelling: assuming arbitrarily that the first vampire appeared in the year 1400, that vampires feed once a month (a 'highly conservative assumption given any Hollywood vampire film'), and that each time a vampire feasts upon a human, their respective populations increase/ decrease by one, a basic geometric progression suggests that vampires would wipe out humans in approximately 2.5 years. There is no way that human birth rates could outpace this, so our continued existence precludes the existence of vampires.

A Norwegian study, however, claims that vampires are real and that the Balkans are especially haunted.[7] Is it possible, the authors ask, to repel vampires with garlic? No vampires were available for study so leeches were used instead, and it turns out that leeches by far prefer a hand smeared in garlic to one without. The authors therefore recommend tight restrictions be placed on the use of garlic in vampire-dense regions.

We can also stop worrying about zombies. The usual zombie paradigm is similar to that of vampires, so the same mathematical

---

[*]   Including 'astronauts currently aboard the international space station . . . Dungeons and Dragons players, men who have read *Fifty Shades of Grey* and other similar beings likely to be unaffected by the apocalypse'.

[†]   'It seems strange to have a supernatural power that only allows you to get around by mimicking human ambulation . . . a very slow and awkward way of moving about in the scheme of things.'

logic applies. However, isolated cases of zombification are apparently possible. In one curious case, Haitian boy Wilfrid Doricent appeared to be dead, but returned from the grave, without memory or effective cognition, having dug himself out. The zombie effects appear to have been caused by a poison brewed by an angry uncle (using the toxin from a puffer fish similar to that used in the Japanese delicacy fugu), while non-fatal brain damage had been caused by the lack of oxygen available in the grave.[8]

Figure 15: Academic Halloween costumes

Academics have taken to Twitter like a duck to Twitter: around one in forty scholars now admits to using the microblogging site.[1] While there is the inevitable scholarly chat and self-promotion, Twitter also acts as something of a virtual water cooler, a place where academics go to build community, have some fun, and let off steam.

I feel smarter just by following the likes of astronomer Katie Mack and The Lit Crit Guy, who have a knack for posting witty and engaging musings on fields I normally wouldn't venture into,[*2] and a few academic superstars have built up a level of fervent popularity that would have been unimaginable before social media.

As with other online communities, academics have created a host of niche parody accounts. Academic Batgirl and Research Mark (Wahlberg) are perennial favourites, and there is an (over)abundance of Angry

---

* @AstroKatie became known to the internet more broadly in 2016 due to her quick comeback to a climate sceptic troll who told her: 'Maybe you should learn some actual SCIENCE... stop listening to the criminals pushing the #GlobalWarming SCAM!' She responded: 'I dunno, man, I already went and got a PhD in astrophysics. Seems like more than that would be overkill at this point.' J.K. Rowling posted a screenshot of the tweet which was liked 165,000 times, doubling Mack's following overnight.

Professor/Overworked Grad Student type accounts. Fake Elsevier does an excellent, albeit sporadic, job of poking fun at the traditional academic publishing model, while others, like Shit My Reviewers Say, lift the lid on the publication process. Even the Oxford comma has its own account.

- Elsevier's new sharing policy allows you to verbally explain your scholarly work to badgers and other woodland creatures.

    Fake Elsevier (@FakeElsevier)

- Call me kinky, but I like to be used.

    Oxford Comma (@IAmOxfordComma)

NEIN

Former Ivy League German professor Eric Jarosinski admits he was initially internet averse. A few years ago a friend introduced him to Twitter. While he didn't get it at first, he followed a bunch of comedians and soon started to see its potential. Then he started *Nein Quarterly* (@NeinQuarterly).

Nein promises a 'Compendium of Utopian Negation' and delivers a unique brand of nihilistic snark and sarcasm. Eric's following has grown to around 150,000 followers, and is supplemented by a weekly column in the prestigious German newspaper *Die Zeit*. Impossible to pigeonhole, Eric says he is simply writing jokes inspired by the terse and astute observations of Karl Kraus, an early 20th-century Austrian writer and satirist, and others.

His pithy musings incorporate word play, puns, contradiction, and are often linked to current affairs or daily life.

- If you need me, I'll be wondering why. Then how. Then for how much longer.

- Youth. Wasted on the wrong demographic.

- The Tickle Me Werner Herzog I got for Christmas only laughs when I tell him the universe isn't utterly indifferent to our pain.

- Every now and then you should step back. Take a look at your life. And keep stepping back.

The constraints of Twitter's 140 characters was a welcome antidote to the frustrations of academic writing. In an interview, Jarosinski told the *Local*, 'It feels so different than the emptiness of a whole page on a laptop and so those constraints for me really brought about the creativity.'[3]

Not only that, but he was good at it. In July 2015, he quit his job at the University of Pennsylvania to develop the alter ego full time. He has now toured the world and published *Nein: A Manifesto*.[4]

SHIT ACADEMICS SAY

Professor Nathan C. Hall is a professor in the learning sciences programme at McGill University. He is also the creator of the wildly successful *Shit Academics Say* (@AcademicsSay).

Initially anonymous, in his revealing interview with the *Chronicle* Hall describes himself as 'A rank-and-file academic with the job of balancing respectable research with acceptable teaching evaluations and sitting on enough committees to not be asked to sit on more committees.'[5] In fact, he is undoubtedly one of the most influential, and funniest, people in academic social media.

Hall started the SAS Twitter account in September 2013, with the inaugural tweet, 'Don't become an academic'. He now has almost a quarter of a million followers.

After a few years in the ivory tower, Hall was feeling fatigued by academic life. As he approached the holy grail of tenure, he started to feel the need to do something a bit different, so he got a Twitter account. 'It's hard to describe the giddy grade-school excitement of jumping into a rapid-fire fray of remarkably creative, clever, and brutally honest tweets from academics around the world', he told the *Chronicle*.

Hall has a wicked sense of humour and his tweets, sent from his phone while working out or waiting for the end of his daughter's ballet class, are a hit. The most popular fall into two categories: snarky quips that are instantly relatable to almost any academic, and amusing riffs on common phrases and clichés.

Favourites of the former kind include:

- I do my best proofreading after I hit send.
- I am away from the office and checking email intermittently. If your email is not urgent, I'll probably still reply. I have a problem.
- Deep down, academics want the same thing as everyone else: acceptance, with minor revisions.

Those in the latter include:

- Give a man a fish, he'll eat for a day. Teach a man to use gender-neutral pronouns and he'll feel uncomfortable with many popular metaphors.
- Choose a discipline you love and you'll never work a day in your life likely because that field isn't hiring.
- Two academics walk into a bar. They bring their own drinks, pay $5,000, and leave feeling proud and ashamed. It's a publishing metaphor.
- If you can't say anything nice.[*]

In a particularly self-reflective tweet he says: 'I'm not procrastinating. I'm actively engaging in the disruption of traditional academic narratives via social media.'

Hall has indeed been doing more than just procrastinating. He started *SAS Confidential*, a blog covering pressing issues in academic life, and has used SAS to recruit thousands of faculty and graduate students into a comprehensive study into the psychological well-being of academics.

---

[*] Say it in a footnote.

THE ACADEMIC TWITTER SUPERHERO

Dr Academic Batgirl is an Associate Academic Superhero and Overall Badass. She spreads scholarly peace and academic love, all the while protecting Gotham from academic posers and offensive grammar.*

Nice cape! Do you wear it in the office?
I'm considering it. Twitter has been witness to heated debate over what 'acceptably dressed' women – and, in particular, professors – should wear. I think if I wore my cape to class, research meetings, and faculty council, it might solve that wardrobe nonsense.

What gave you the idea of developing an academic alter ego?
Academic Batgirl is a superhero in two places where gender is a big deal: the ivory tower and the jungles of social media. Full-time male faculty members still outnumber women by nearly 20%, and, among other inane gaps, gender biases have been shown to exist in the perception of quality in scientific studies. When I first joined Twitter, there were no female academic meme-makers. My pal *Research Mark* needed a female counterpart. To my knowledge, there are still no other female academics who make memes for scholarly consumption. Bam!

Why Twitter?
There's a particular challenge and joy to being witty and interesting in 140

---

* In real life, she holds a PhD from Cambridge University and is an Associate Professor at a top-flight research university in North America.

characters. Plus, there are more academics with a Twitter presence than on other social media platforms – Twitter is clearly the happening place for academics.

Why Batgirl?
She's a librarian by day and a badass, crime fightin' ass-kicker by night (kind of like most academic women), so the choice was obvious. My Batgirl persona is also a hat-tip to my brother – when he was a kid, his teachers feared that he truly believed himself to be Batman, so we've got a Bat theme in my family.

You left for a while and Gotham mourned. Now you're back! Where did you go?
I had been on Twitter for about a year and had over 5,000 followers. My presence, in addition to making clever memes, included offering support to early career academics, advice to writers, and creating a sense of academic community. Then, I happened upon a real-life Joker. He was a full professor with a razor-sharp mind, enviable intellect, and remarkable ability to quantify any data by any means possible.

However, he was ridiculously controlling and didn't like that I was my own woman. He told me, right to my face, 'This Twitter account is nothing to be proud of.' 'You want me to tell people that you run this Academic Batgirl account? That's embarrassing.' 'No real academic would need Twitter to help with their career.' 'People who use social media are less trustworthy.' And, 'I would be a lot happier if you just quit this whole Twitter thing.' Like many bullied people, I gave in.

I'm sorry to hear that. What made you come back?
The need for academic superheroes is very real. So real, in fact, that I couldn't even satisfy the need in my own offline academic life. I had 'outed' myself to two people on Twitter, and told them what had happened in real life that forced me to disappear. These friends contacted me several times to let me know that people were asking where I went, why, and when I'd

be back. These friends really did call me back to my senses. I mustered up the guts to tell the critical Academic Joker where to go. And it was awesome. Pow!

What kinds of people do you follow and why?
I follow some mind-blowingly interesting and intelligent scholars. I've connected with a dog psychology scholar, a chemist from my hometown, and a legal studies scholar. I follow cool people who study volcanoes, wolves, botany, surgery, palliative care, and queer culture. I have no training in these areas and I don't publish in the journals that they do. I follow these folks because they are good people doing meaningful research, and learning is fun. In addition to following, I've become legit friends with some of these super cool folks.

Favourite hashtag?
#GetYourManuscriptOut. Somehow I've become one of its biggest proponents, along with Raul Pacheco-Vega and Steve Shaw.* I support this hashtag because it has sincerely helped me. I struggle in that I get distracted by new, shiny research projects, and I sometimes think I'll ditch the manuscript I'm working on to get started on something different.

A lot of academics get bored easily, and many of us suffer from thinking that the next research project will be more fun, more successful, or maybe just easier. The #GetYourManuscriptOut hashtag helps to build collective support for finishing what you started (ooh! I sort of quoted Van Halen there).

Most popular tweet?
I made a meme featuring Yvonne Craig as Batgirl, and she's wearing a stern expression. The text reads:

---

* I too follow Raul (@raulpacheco) and Steve (@shawpsych), both of whom are great for a motivation boost.

> *'I AM AN ACADEMIC. This means that I live and*
> *work in a fantasy world in which everything is proofread*
> *(twice), and no one believes anything he/she hears or*
> *sees without consulting the data. THANK YOU FOR*
> *UNDERSTANDING.'*

I think that one was so popular because it's not discipline-specific, and it captures the meticulousness of academic thought and lifestyle. People must have seen themselves in that meme – the serious expression with a rather self-deprecating sentiment was funny because it read like an academic PSA. As in, 'I know I'm ferocious about loving data and being rigorous in all inquiry, but you love me anyway, don't you?'

Favourite onomatopoeia(s)?
Bam! Pow! Zowie!

THE DARK SIDE OF ACADEMIC TWITTER

If these people (and anthropomorphised punctuation) represent the best of academic Twitter, the now-defunct @GradElitism represented the worst of it. The account had managed to attract 40,000 followers by reposting others' jokes without attribution (i.e. plagiarising). A self-appointed watchdog (who later turned out to be Nathan Hall) sprang into action, calling out the plagiarism and getting the offending account closed in a matter of weeks. This brief campaign was no doubt buoyed by the news that Twitter had started clamping down on joke theft.[6]

Some of the darker, spammier corners of academic Twitter don't make any sense to me at all. For example, there's a cluster of profiles that look like student accounts, but post nothing but a never-ending stream of tweets advertising university courses (one that I see all the time posts one tweet every five minutes, over 300,000 in total). They are then instantly retweeted by 20–100 similarly pointless accounts. Presumably this is a marketing ploy, but it would take a lot more than an onslaught of overwhelmingly bland tweets to convince me to take an 'Introduction to Mathematics' course.

A recent addition to the list of questionable Twitter enterprises is *Real Peer Review*. Run by a group of anonymous academics, the account aims to pick out papers that they believe are lacking in intellectual rigour or value. The group argues that 'such laughably broken "research" is a natural consequence of any sufficiently isolated and ideologically homogenous community' and takes a 'sunlight is the best disinfectant' approach to rectifying it.[7]

In particular they are critical of the creation of 'journals focusing on an extremely narrow and insular circle of readers and authors who engage in a kind of obscurantist pseudo-intellectual mutual masturbation (often with some degree of public funding) with absolutely no measurable or even coherently expressible benefit to the field'.

The account has grown rapidly in popularity, especially as news spread that an earlier incarnation was shut down when the original founder received threats from enraged academics.

I have mixed feelings about this. On the one hand, I instinctively baulk at the idea of an anonymous account singling out particular papers for ridicule. On the other hand, some of the papers highlighted by this rogue band of academics are truly confounding.

Title	Slightly satirical one-sentence summary	Quote
'Sleeping Around, With, and Through Time: An Autoethnographic Rendering of a Good Night's Slumber'[8]	Academic sleeps at her two houses, in a plane, and in a hotel, spends 11 pages talking about it.	'I turn. Art turns. Between us, Buddha turns, then hops over me so she doesn't get squashed by two human bodies rolling together. I stretch out my legs, disturbing Zen who is snoring at the bottom of the bed... Art snuggles in close to me, his chest and knees pressed against my back and legs...All is well here in our king-sized, platform bed; together we perform the twists and turns of our sleeping ritual, escaping from the tensions and noise of the outside world.'
'Club Carib: a geo-ethnography of seduction in a Lisbon dancing bar'[9]	Academics go clubbing three nights a week for two years, find that Lisbon's nightlife has a '(hetero) normative and patriarchal character'.	'Some tourists, Erasmus students and young Portuguese students drink in order to socialize by sharing time, space and experiences with their peers. Others drink just to escape from their harsh individual realities. Many hope for an unforgettable night (and perhaps another in the future).'

'"I'm Here to Do Business. I'm Not Here to Play Games." Work, Consumption, and Masculinity in Storage Wars'[10]	Academics watch Storage Wars* find that it 'helps mediate the putative crisis in American masculinity'.	'By emphasizing the economic benefits (i.e., masculine) of bidders' quest for thrift rather than the hedonic and relational benefits (i.e., feminine), Storage Wars suggests auction bidding allows for the ritual transformation of spending – a frivolous and wasteful act – into a productive act.'
'The Perilous Whiteness of Pumpkins'[11]	Academic buys pumpkin spice lattes, realises they are oppressive symbols of white privilege.	'To explore race, culture, and food, we turn to three recent moments in the narrative of pumpkins' whiteness: the pumpkin spice flavor industry; the Internet phenomenon, "Decorative Gourd Season," and lifestyle magazines' fall embrace of class-aspirational pumpkins; and the working-class reality television Punkin Chunkin contests.'†
'Group Sex as Play: Rules and Transgression in Shared Non-monogamy'[12]	Academics hang out at swingers parties, find that swingers have a lot of fun, but also a lot of rules.	'I'm sitting on a couch, watching a gorgeous man being fisted on a sling. The woman leaning next to me lets out a long, pleased sigh: a lover has just entered her, unannounced, from behind. The researcher in me immediately thinks "she did not have time to indicate consent," then remembers that this is not the first time I've watched them have sex tonight. They have obviously reached an agreement.'

* A US 'reality' show about abandoned storage units and the people that make a living buying them blind at auction

† i.e. pumpkin throwing

#HASHTAGS

For those blissfully unaware of the machinations of the Twittersphere, hashtags are used to collate tweets on a specific subject.[*][13] As such they have proven to be a great tool for community building, with regulars such as #PhDchat, #AcWri, and Raul Pacheco-Vega's #ScholarSunday providing opportunities for academics to interact and learn from each other. Others, such as #AcademicsWithCats and #AcademicsWithBeer cater to extra-curricular interests. The recently coined #AcaDowntime encourages the academic community to take time away from work, and a skim through the feed reveals that academics are an interesting and active bunch. Hashtags are also used to play games and joke around, which is where the real fun begins.[14]

#AcademicsWithCats

#AcademicsWithCats was one of my earliest forays into hashtags, and I am proud to say that it is now a staple of the academic Twittersphere. These days the feed is mostly pictures of cats engaged in decidedly non-academic activities, but the glory days produced some fantastic pictures of cats reading Nietzsche, correcting essays with a red pen in paw, and hammering out essays on laptops.

The hashtag spawned the annual Academics With Cats Awards, which provides a bit of light relief toward the end of the year. The awards have been covered by the higher education supplements of the *Guardian* and *The Times*[15] – in 2016, around 500 academics entered and over 2,000 cast a vote for their favourite feline.

---

[*] One of the rabbit holes I went down while writing this section was trying to discover the origin of the hashtag, and then the hash symbol itself. This is a fascinating story (honest), beautifully told in an episode of the podcast, *99% Invisible*. If you've been meaning to figure out what these new-fangled podcasts are all about, I'd highly recommend starting with *99% Invisible*, which is worth listening to just to hear presenter Roman Mars's voice.

#BadAdviceForYoungAcademics

Oscar Wilde is famously quoted as saying, 'sarcasm is the lowest form of wit'.* When I saw a hashtag being used to give advice to young academics, my first reaction was to join in with sarcasm. #BadAdviceForYoungAcademics was born and academics in their thousands chimed in to offer their un-advice. The sarcastic advice was much more fun (and probably just as helpful).

Writing:[16]

- Write your thesis in comic sans.
- Grammar be optional, it are what you says that mattering not how you say it.
- Just submit the paper. You can fix the bad writing and bogus results later.

Publish or perish:[17]

- Third author of eight is really an important position, especially when you did all the work and wrote the paper.
- Don't publish during your PhD, there's plenty of time for that later.
- Reviewers will respect you for challenging their critique and pointing out their idiocy.

Presentations:[18]

- No need to practise your presentations, just wing it. You'll be fine!
- Make sure your Prezis feature lots of movement.
- Moonwalk to the front before a presentation. It's good to maintain eye contact with the audience from the outset.

---

* My Mom always used to tell me that Wilde went on to say that sarcasm is the highest form of intelligence, though this part is usually omitted from the quote.

Career advice:[19]

- Don't worry. Funding is plentiful.
- Trust that a tenure track position awaits you.
- Lots of older Profs will be retiring in the next few years.
- Now is the perfect time to go into academia. Universities are desperately searching for people to fill positions.

#RuinADateWithAnAcademicInFiveWords

This simple twist on the 'ruin a date' game gave us a fascinating insight into the academic psyche on one of life's precious pleasures – love and romance.

To spoil a date with an academic, you can say something stupid like:[20]

- Is that all you've published?
- Oh, you're not tenure track?
- So people read your articles?
- What is the practical application?

A sure-fire mood-killer is expressing admiration of/interest in any of the following: Fox News, astrology, homeopathy, Ayn Rand, or the History Channel. Disavowal of reading, evolution, coffee, and the Oxford comma might also end badly, and remember: hell hath no fury like an academic who's been asked if they get summers off.

Other things likely to end with a drink in your face are asking how your date's PhD thesis is going or when it will be finished, telling them that said PhD does not make them a *real* doctor, and completely misunderstanding their field ('Astronomy? Cool, I'm a Virgo!').

The majority of the tweets assumed the date-ruiner to be the non-academic party, but plenty of people realised that having two academics at the table could be the real recipe for dating disaster:[21]

- [your discipline] is really just [my discipline].

- I applied for the same funding.
- Yeah, I was Reviewer 3.
- Let's meet during office hours.

#FailAPhdInThreeWords

While it might take five words to ruin a date with an academic, the Twittersphere proved that a PhD can be ruined in just three:[22]

- Computer dead. Backupless.
- Ethics permission expired.
- What primary sources?
- *Mein Kampf* reconsidered.
- Dog ate it.
- Supervisor sex tape.
- Cf. Mum, Your.
- It was aliens.

#ScienceAMovieQuote

There is an excellent scene in *The Martian* where, after realising he has been left alone to eke out an existence on Mars, Matt Damon's character says emphatically: 'I'm going to have to science the shit out of this.' Around the same time, the science folk of Twitter decided to science the shit out of movies in a beautiful marriage of science and movie geekery:[23]

- 'I love the smell of null hypothesis rejection in the morning.'
- 'I sequence dead people.'
- 'We're going to need to write a grant for a bigger boat.'
- 'I'm just a girl, standing in front of a rat, asking him to press a lever.'
- 'Say hello to my little trend.'

#AcademicForecast

People have, on occasion, asked how a particular hashtag came about. I have often wondered the same of others' hashtag creations (#PhDAsExistentialCrucible, anyone?) but usually struggle to remember what the thinking was behind my own.

One day I started out trying to tackle some 'minor revisions'. An hour in, I realised that so-called minor revisions are rarely minor. Admittedly, some of the reviewer's comments were easily answered (e.g. I had neglected to capitalise the word 'Tuna'), while others, innocuous at first glance, were Pandora's boxes of academic pain.

I turned to Twitter to procrastinate but my feed was overflowing with snarky tweets from internet pedants. Faced with pedantry from Reviewer 2 or pedantry on Twitter, I made a forecast: '90% chance of pedantry on Twitter, otherwise acceptable with minor revisions.'

I quite liked the idea of an academic day being summed up by a slightly sarcastic weather forecast, and figured that others may wish to join me. They did:[24]

- 'Outlook uncertain. Copyright handed over to publisher, peer review highly likely, acceptance rate 26%. Rejection expected.'

- 'Strong, gusty modelling until 13:00, followed by brief exposure to daylight, then heavy spreadsheets.'

- 'Heavy morning fog, lifting as caffeine levels increase. High chance of distraction with possible tweeting.'

#AcademicNovel[25]

- Harry Potter and the Half-Written Thesis
- Harry Potter and the University of Phoenix
- For Whom the Bell Curve Tolls
- Where the Tired Things Are

- The Lord of the Files
- 20 Thousand Leagues of Self-Citation
- The Grades of Wrath
- Fear and Loathing on the Tenure Track
- The Curious Incident of the Grant in the Pipeline
- The Winter of Our Job Market Discontent

OVERHEARD ON TWITTER

Students say the funniest things on Twitter, apparently unaware or unconcerned by the highly public nature of their musings. They brag about plagiarism, trash-talk their tutors, and laugh about skipping class. Occasionally more amusing (or concerning) than students' own grumblings are some of the things they quote their professors as saying:

- 'I drink like a fish. I can drink you all under the table!'

- 'Papers should be like a woman's skirt. Short enough to be interesting but long enough to cover the subject.'*♀

- 'My music professor makes us stay after class and play Twister with him to make up attendance. Dead serious. I find a problem with this, no?'

---

*   The sheer frequency of this one is astounding and concerning in equal measure.

OBSCURE INTERLUDE

———— ∞∞∞ ————

LOVE AND ROMANCE

Tired and unloved? Working an 80-hour week with no time for dating? Put down your red pen, back away from the UCLA Loneliness Scale,[1] and read on.

Academics have conducted an awesome array of research on love and romance. Some of this is pretty common sense stuff: you are better off single than in a dysfunctional relationship,[2] but unhealthy relationships are easier to fall into once you have been alone for a while (because we settle for less when we are lonely).[*3]

Rate My Professors could help narrow down the field. Cute academics abound in the language department, while if it is intelligence you seek, philosophy and political science is where you shall find. Steer clear of the music school if you are not a fan of elbow patches and tweed.

Science says that you should get into the sack as often as possible (more sex means fewer colds,[4] not to mention that it is good exercise). There is a vast sexology literature that can help, but the best nugget of amorous advice is this: wear socks. A study on the female orgasm found that only half of participants were able to achieve orgasm without socks, but this jumped to 80% with them.[5] Apparently warm and cosy feet calm the amygdala and prefrontal cortex – the brain regions responsible for anxiety and fear.

---

* Best not start dating during your PhD, then.

Even if you do find a mate, love might still get you in the end. Being in a relationship causes weight gain,[*6] and the medical literature reports on many cases of 'Broken Heart Syndrome'. One case report discusses a 70-year-old woman with no prior heart problems who collapsed in hospital after being informed that her husband of 45 years had died.[7] While this is the stuff of urban legend, the jury is still out on the causal link.[8]

#ElsevierValentines[9]

- Roses are red, Violets are blue, Copyright is ours.

- Roses are red, Dollars are green, Scientists' free work, Keeps our profits obscene.

- Roses are red, Violets are blue, Please give me your heart, So I can sell it back to you.

Figure 16: Academic Valentine

---

* Though this is according to research commissioned by a dieting company and reported by the *Daily Mail*. I'll say no more.

CONFERENCES

An academic conference can be anything from a small and collegial meeting of minds in a quiet campus block, to a grandiose affair involving thousands of participants and spanning multiple days and venues.

Irrespective of size, the unifying certainty of academic conferences is the ubiquitous panel discussion. At some point, presumably at a conference on conferences, it was decided that the standard format for an academic conference would be the panel discussion. ISO standard 3103 defines an academic panel as a parade of three to four speakers taking it in turns to read from their PowerPoint presentations, followed by questions from the otherwise bored-to-tears audience.[*]

Custom dictates that the majority of panels feature only male speakers;[♀1] that slides should be overfilled, illegible, and written by the

---

[*] There isn't really an international standard for a conference panel. There is, however, a ISO standard for wooden panels used to test paint (ISO/TC 89), which is more or less the same thing, given that most speakers are wooden and the panels are like watching paint dry. ISO 3103 cited above is, in fact, the ISO standard for brewing tea, which won the Ig Nobel for Literature in 1999. While the standard does factor in water hardness and the prohibition on reboiling, it makes no recommendation regarding pre-warming of the teapot. ISO standards are reviewed every five years and I shall be writing to the ISO Technical Committee on Food's Sub-committee on Tea to correct this oversight just as soon as I have finished writing this book (if I ever finish writing this book).

speaker on the way to the conference; and that audience questions should actually be long-winded comments unrelated to the speaker's presentation (and/or thinly veiled resentment at the questioner not having been invited to sit on the panel themselves). It is also customary for the chair of each panel to abdicate all responsibility for timekeeping, such that the coffee breaks and lunchtime (the bit that I find most interesting and productive) get condensed into a vanishingly small time slot. There must be a better way,* but, for the moment, the panel reigns supreme.

As the conference itself is unlikely to be a life-changing experience, there is only one question to ask yourself before deciding whether to go: Where is it? This is no doubt why a great many conferences seem to take place in holiday spots that seem otherwise unrelated to the conference topic. Why go to a symposium to give a presentation when you can go to a skiposium for a presencation?[2]

If you are looking for a grant-funded getaway, the Academic Organization for Advancement of Strategic and International Studies (OASIS) may be a good place to start. Its website says that it is an 'Association of dedicated professionals, who willingly devote their capabilities in an ethical way for the betterment of our local communities and the society in general.[3] Yet the organisation's name,† logo (palm trees), and website (which opens with a picture of some generic beach city) belies this mission. The organisation supposedly publishes a few open access journals,‡ and organised six conferences in 2015: Miami Beach, Key West, Paris, Bangkok, Orlando, and Las Vegas.

---

* Thinking out loud: conference speed dating, papers presented through mime, presentations tweeted using only lolcats …

† The name has been changed a few times. Most recently it was called the Institute of Strategic and International Studies, but presumably changed that when the acronym suddenly became untenable.

‡ All with an extremely broad scope. The website claims that papers are 'double-blind peer-refereed' in 3–5 weeks, a timeframe that seems highly unlikely. The journals are not open access. In fact, you can't seem to pay for access – even the table of contents is impossible to look at.

If resorts and gambling havens aren't your preference, you can always find conferences in more compelling locations. I asked academic Twitter where the best or weirdest places they'd been to conferences were, and the range of responses had something for all tastes: [4]

- 'Sorrento or Prague for the sweetest, most beautiful. Fargo, ND for the . . . opposite'.*
- Halfway up an active volcano.
- The Tower of London (for a conference on Renaissance imprisonment).
- Boiling Springs, North Carolina ('was pretty odd').
- In an Edwardian swimming pool (presumably empty).
- Timberline Lodge, Mount Hood, i.e. Overlook Hotel from *The Shining*.
- A converted Benedictine monastery ('Definitely felt like we were getting our Umberto Eco on').
- A half-built hotel on St Kitts that had its electricity cut mid-conference due to non-payment of their bill.

Some conference settings seem better suited to a Dalí-esque silent film. Sarah Young from University College London recounted her visit to the Annual Conference of the Slovenian Comparative Literature Association.[5] The conference was held at Lipica Stud Farm in Slovenia, and the conference sessions were held in old stables surrounded by paddocks of dancing horses. The participants stayed in a desolate hotel-cum-casino on the Slovenian–Italian border. Young admits that it was a struggle to concentrate and that she was left with little recollection of some of the papers. Daniel Jagger, also from UCL, recounted the 2010 midwinter meeting of the Association for Research in Otolaryngology, which took place at Disneyland.[6] 'Goofy & Snow White waving at scientists in the

---

* In a stroke of social media genius, the Twitter account for the city of Fargo replied: 'We'll take that as a compliment!'

lobby was weird . . . We booked a taxi to the theme park [and] arrived in a stretch Humvee. With internal disco lights.'

SHODDY CONFERENCES

The same predatory publishers that spam our inboxes offering publications now do conferences too, trying to pass them off as legitimate academic gatherings to extract money from researchers.

I receive a handful of such requests a week, excluding those that are so spammy that they are binned by the junk filter before they even reach me. Sometimes these conferences have names that are almost indistinguishable from the names of real conferences, and often boast big names as speakers and organisers, even though these people haven't actually agreed to participate.

Gina Kolata, writing about this parallel world of pseudo-symposia in the *New York Times*, highlights the example of the unfortunate scientists who paid to present at *Entomology-2013*, thinking they were going to *Entomology 2013*.[7] 'I think we were duped,' said one of the attendees in an email to the Entomological Society. They just have to hope that the department heads reading their résumés later on also fail to spot that tricky hyphen.

Disgruntled at having been taken in by a dodgy conference, one blogger shared their experience.[8] The name and website of the conference created a grand impression: 24 conference organisers including high-profile scientists; 11 thematic tracks; and pictures of a big conference room. But the cracks were starting to show before the conference had even kicked off. The participants received scant information regarding logistics, and the 11 conference 'tracks' had been condensed into a single 'stream'.* The resulting programme was crammed so full that there were few breaks. In spite of these early warning signs, the author of this exposé says that he 'really wasn't ready for the shambles that was to come'.

---

*   I often wonder if all this talk of 'tracks' and 'streams' in conference programmes is the result of our subconscious desire to be frolicking in the woods instead of sitting in a conference room.

Highlights included:

- A conference pack consisting mostly of advertising for other conference activities (attendee list not included);
- Just two 15-minute coffee breaks and 40 minutes for lunch in a nine-hour day;
- A tiny venue, because the room shown in the brochure had been divided and the other half was being used to host another of the company's conferences;
- A 30-minute opening 'ceremony' – in fact an awkward five-minute introduction from one of the keynotes who had been hastily ushered into the role; and
- Speakers going AWOL, with the organisers having no knowledge of their whereabouts ('Before each talk, there was a hopeful appeal to the audience for the speaker to come forth and show themselves – or, as in a few cases, not.')

There were apparently some great scientific presentations, though the disappointing overall experience was not improved by the overzealous Certificate of Recognition given to participants, in which the organisers 'enjoy special privilege to felicitate [name] for his/her phenomenal and worthy oral presentation'. To add insult to injury, they added this academic to its list of Executive Editors (without asking, of course).

This appears to be far from an isolated incident, though few are brave enough to recount their experiences in such detail – as the author notes, it can be a bit embarrassing to admit that you were duped in this way.

It's not just sham conferences that can be shoddy – sometimes the real deal can be just as underwhelming. So common are such occurrences that some academics got together to make a bingo card generator and turn it into a game.[9] Squares include: overenthusiastic air-conditioning, coffee that breaches the Geneva Conventions, food issues, and misspelled names on conference tags.

These conferences represent the nightmare. My dream is a conference with hot tubs, popcorn machines, and WWE-style intro videos for keynote speakers.[10] Failing that, I'd be happy with decent coffee, free WiFi, and the abolition of panels.

KIMPOSIUM

I've never been sure what exactly Kim Kardashian does (and I honestly haven't had the inclination to find out), yet she crops up surprisingly often in academia. In November 2014 Brunel University hosted a symposium on the Kardashians (a 'Kimposium').[11]

While I am yet to be convinced of the cultural significance of Kim's internet-breaking bottom, the famous family, it is argued, are influencing discussions of race, feminism, and beauty. Conference organiser Meredith Jones, reader in sociology and cultural studies at Brunel, told *Times Higher Education*:[12]

> *You may love them or hate them, but the Kardashian family must be examined ... They may be vacuous and bland when they open their mouths, but they are also very powerful. It is silly to think this subject is not worthy of academics' attention.*

The day-long meeting included a range of talks, including 'Kim Kardashian as the embodiment of the networked-image', and 'Media-Bodies: what Kim Kardashian's vulva can teach us about contemporary life'.[12]

CONFERENCE ETIQUETTE

'More of a comment than a question,' the academic says, rising assuredly from their seat and launching into a lengthy exposition of their own recent publication and/or metaphorically ripping the speaker's paper to pieces.

If you've ever attended an academic conference, this scene will likely be familiar. The presentation portion of the proceedings has finished, the microphone is passed to the floor, and an enthusiastic audience member is yearning to seize the spotlight (generally prefacing their remarks with an unnecessarily long autobiographical introduction).

Why, as Stacey Patton from the *Chronicle* puts it, do academics 'risk coming off like jackasses at conference Q&A sessions?'[14] Anna Post (great-great-granddaughter of famous etiquette author Emily Post) reckons those who like to show off by highlighting key lines from their CV or slipping in a few Latin or French phrases into their remarks are simply insecure: 'People who do that are usually not the most popular people in the room,' she opines.[15] Of course not: the most popular people in the room are those with the WiFi password.

Other unbecoming behaviours commonly seen at conferences include the inevitable skirmishes for scarce plug sockets and participants showing up visibly hungover.* There is always one attendee who rolls in late, bumbles to the front row and immediately begins whispering audibly in the ear of the poor person next to them. Then their phone starts vibrating and they scramble to answer it, before scurrying to the back of the room to conduct their conversation, again in not-so-hushed tones. Always try to identify this person early on – they will help bring you victory in conference bingo.

---

* I confess to having done both on multiple occasions, and both simultaneously on at least one occasion.

CONFERENCE BINGO!

'Sorry if you can't read this at the back.'	Horrifying gender ratio	Prezi	Lecherous academic	Unnecessary rhetorical flourishes
Overenthusiastic air conditioning	Presenter really proud for having reinvented the wheel	Death by PowerPoint	Skiving a session	Popular academic is scheduled in a tiny room
'I'll try to be brief.'	Obtuse handouts	FREE SQUARE	Conference held over major holiday when hotel costs soar	Last minute change of programme
Gratuitous and unnecessary use of French critical theory	'Thank you, I enjoyed your talk very much.' Proceeds to utterly destroy talk	Plenary is n^{th} permutation of paper toured over the last two years.	Pie chart	Poster assembled as a collage of A4 printouts
'Anybody got a DVI to VGA adaptor for my Mac?'	Accessibility issues. Any and all	Extensive use of buzzwords	Clash of the academic alpha males	'XYZ doesn't really need an introduction...' for plenary speaker

Fellow presenter takes drink from your water glass	Unrealistic number of slides	FREE SQUARE	Every bullet. On every slide. Flies in. Separately.	Awkward silence
Presenter doesn't show up	Chair purposefully mispronounces speaker's name	Clip art	'Sorry, I'm a Mac [or PC] person,' said by the presenter who can't get the slideshow working	Alcohol reception with insufficient alcohol
Leading academic skives entire day	Question is completely about speaker's own project	Infomercial masquerading as panel presentation	Conference scheduled for predictably inhospitable climate	Snoozing academic
'In a longer/earlier/different version of this...'	Weak chairing means that last speaker of four, who has travelled furthest to get there, gets five mins instead of twenty	Conference theme is curiously absent from the programme	Seasoned academic asks junior presenter a mercy question	Food issues. Any and all
Presentation in large lecture theatre attracts <15 attendees	Poorly formatted abstract book	Long winded debate between two people with no room for anyone else	Someone shows up visibly hungover	Live tweeting

OBSCURE INTERLUDE

⸻⸻⸻

CAMPUS HIJINKS

At the centre of the campus of the University of Southern California sits a stately statue of the school's unofficial mascot, 'Tommy Trojan'. In 1958, a group of students conspired to coat Tommy in manure and rented a helicopter to dump their noxious cargo. As they attempted to disperse the manure it was drawn into the helicopter's rotor blades, spraying the students with a taste of their own medicine.

That same year, Peter Davey of Cambridge University started the trend of sticking cars on campus rooftops. Following months of planning, reams of calculations, and help from students who volunteered to surreptitiously erect scaffolding, he hoisted an Austin Seven 70 feet to the top of the Senate House. It took a week to get the car down afterwards. In 1994 some MIT students followed suit, putting a fake campus police car atop the dome on Building Ten and issuing it with a parking ticket.

Perhaps feeling that cars on rooftops had become passé, students at Carleton College temporarily transformed the university's observatory into a huge replica of R2-D2. The swivelling of the telescope made it the perfect medium, and the likeness came complete with all the robotic beeps of the original.[1]

Some of the biggest and best university pranks have been pulled off during college football games, which are a big deal in the US.[*2] The 1961 Rosebowl was watched live by 100,000 spectators, and by millions on TV (by comparison, Wembley Stadium has space for 86,000 spectators).[†] They were shocked when fans held up cards which, taken together, read 'CALTECH'. Tiny Caltech, the California Institute of Technology, is better known for science than sports, and were not even playing in the match. Crafty Caltech students had convinced a cheerleader that they were journalists, allowing them to sneak into the cheerleaders' hotel rooms, and switch the cards and instructions for the fan displays. In 2004, two Yale seniors went one step further: they gathered twenty friends, costumed as the fictional 'Harvard Pep Squad', waltzed into Harvard's stadium, and convinced 2,000 unsuspecting fans to unwittingly spell out the words 'WE SUCK'.

Campus pranks have made it into the classroom too. In 1927, Georgia Tech student William Edgar Smith received an extra enrolment form, so he filled one out for the imaginary George P. Burdell. Smith completed coursework for his fictitious friend, earning him a real degree. Burdell has since become the stuff of university legend, earning many additional degrees and being admitted as a member of a range of clubs and societies. When Barack Obama spoke at the university, he joked that George was meant to be introducing him but was nowhere to be found.

---

* Not just culturally, but also financially. In an attempt to calculate the value of teams, Ryan Brewer from Indiana University-Purdue University Columbus analysed the revenues and expenses of each football programme, then made cash-flow adjustments, risk assessments and growth projections to calculate what a college football team would be worth on the open market. He estimated the value of the top ten most valuable teams in 2015 to be about $7 billion.

† Astute football fans may note that Wembley's maximum capacity is 90,000, but for some reason 4,000 fewer seats are available when it hosts an American football game.

Campus Police Reports, Brigham Young University
25 September–1 October, 2015[3]

Sept. 27 – 'University Police received a call about a transient in the Life Science Building at 10.31 p.m. The transient turned out to be a student who fell asleep while studying.'

Sept. 30 – 'Around 8 a.m., Provo Police dispatch received a call about a moose in the area of 1450 E Oak Cliff Drive that was heading west towards Wasatch Elementary School, according to the Provo Police Facebook page. The responding officers were able to corral the moose in a nearby LDS Church parking lot. When the Utah Division of Wildlife officers arrived, the moose was subdued with a tranquilizer gun. The moose was released back into the wild.'

Oct. 1 – 'A female student purchased $40 worth of food for General Conference weekend and stored it in a communal refrigerator in the basement of Hinkley Hall in Helaman Halls. When she returned the food was gone, and University Police believe it was most likely consumed.'

Oct. 1 – 'A group of students playing hide and seek in the Harris Fine Arts Center at 11 p.m. caused a faculty member to call the University Police. The police arrived but were not able to find any of the students.'

ACADEMIC ANIMALS

Animals are all over academia, from long-suffering lab rats to levitating frogs, and many an Ig Nobel has been won on the strength of an amusing animal study, including:

- 'Walking Like Dinosaurs: Chickens with Artificial Tails Provide Clues about Non-Avian Theropod Locomotion': researchers attached prosthetic tails to chickens in a bid to understand how dinosaurs walked.[1]

- 'Dogs are Sensitive to Small Variations of the Earth's Magnetic Field': found that when dogs go to the loo, they prefer to align themselves with the Earth's north–south magnetic field.[2]

- 'Chicken Plucking as Measure of Tornado Wind Speed': proposed that tornado speed be measured by the speed required to blow all the feathers off a chicken.[3]

- 'Dung Beetles Use the Milky Way for Orientation': discovered that when dung beetles get lost, they can find their way home by looking up at the Milky Way.[4]

- 'Are Cows More Likely to Lie Down the Longer They Stand?': found that it is more likely that a cow will soon stand up after it has been lying down for a long time, but that once it stands up, you can't easily predict how long until it lies back down.[5]

A surprising number of cats and dogs have also been bestowed with degrees or appeared as authors on peer-reviewed papers.

Table 6: Cats and dogs with academic qualifications

Name	Animal	Year	Degree Awarded/ Institution	Notes
Zoe D. Katze	Cat	2001	Hypnotherapy certifications	Zoe received a handful of different certifications ('Not bad for a cat who's not even purebred').[6]
Colby Nolan	Cat	2004	MBA from Trinity Southern University	Cat of Pennsylvania Deputy Attorney General, who paid $299 as part of an exposé.[7] Resulted in a fraud lawsuit.
Henrietta	Cat	2004	Diploma in nutrition from the American Association of Nutritional Consultants	Science Journalist Ben Goldacre's cat. Obtained as part of an investigation into the qualifications claimed by a famous TV nutritionist. ('A particular honour since dear, sweet, little Hettie died about a year ago.')[8]

Sonny	Dog	2007	Medical diploma from Ashwood University	Sonny belonged to an Australian comedian and obtained his degrees as part of a skit on *The Chaser's War on Everything*. The 'work experience' section of Sonny's application to the university included 'significant proctology experience sniffing other dogs' bums'.[9]
Lulu	Dog	2010	Law degree from Concordia College	Mark Howard, a member of the legal team for BskyB during a lawsuit, obtained a degree for his dog from the same alma mater as the defendant.[10]
Pete	Dog	2013	MBA	Pete received a master's degree in just four days, for £4,500. Newsnight reported that Pete (named Peter Smith on his fake CV) was offered the degree based on his fictitious work experience and undergraduate degree.[11]

CATS

If #AcademicsWithCats has taught us anything, it is that academics, like everyone else with an internet connection, love cats.[*][12] But the academic–cat relationship predates the social media era by hundreds of years. Emir Filipović from the University of Sarajevo was trawling through the Dubrovnik State Archives when he stumbled upon a medieval Italian manuscript (dated 1445) marked clearly with four paw prints.[13]

Figure 17: Paw prints on medieval manuscript

It could have been worse. Around 1420, one scribe found a page of his hard work ruined by a cat that had urinated on his book. Leaving the rest of the page empty, and adding a picture of a cat (that looks like a donkey), he wrote the following:

> *Here is nothing missing, but a cat urinated on this during a certain night. Cursed be the pesty cat that urinated over this book during the night in Deventer and because of it many*

---

[*] Indeed, there are even papers in the academic literature trying to work out why exactly cats seem to resonate so intensely with internauts.

> *other cats too. And beware well not to leave open books at*
> *night where cats can come.*

Though occasionally ruining manuscripts, cats undoubtedly saved a great deal of invaluable works by hunting mice that would have otherwise had a field day feasting on the paper. Others, like Jordan the library cat, have taken a less ambitious approach to academic life. Jordan's home is the Edinburgh University friary, but he hangs out in the library, where students fawn over him as he sleeps in his favourite turquoise chair. He has his own Facebook page and the library has even issued him a library card.

One curious cat has outshone all other academic animals. F.D.C. Willard has published as a co-author and, incredibly, the sole author of papers in the field of low temperature physics.[14]

When American physicist and mathematician Jack Hetherington was told that he needed to eliminate the use of the royal 'we' in a paper, he was reluctant to retype the entire manuscript (this was in the days of the typewriter, so rewording the paper would have been a considerable undertaking). To save time, he simply added his cat as a co-author. Concerned that colleagues would recognise Chester's name, he concocted a pen name: F.D. for *Felis domesticus*, C for Chester, and Willard after the cat that sired him. The joint paper was published in *Physical Review Letters* in 1975 and has been cited about 70 times.

When his complimentary printed copies arrived, Hetherington inked Chester's paw, signed a few, and sent them to friends. One of the copies found its way to a colleague who later recounted that a junior physicist on a conference organising committee proposed inviting Willard to present the paper because 'he never gets invited anywhere'.[15] Hetherington's colleague showed the committee his signed copy of the paper, whereupon everyone in the room agreed that the paper appeared to have been signed by a cat. Neither Willard nor Hetherington was invited.

'Shortly thereafter a visitor to [the university] asked to talk to me, and since I was unavailable asked to talk with Willard', Hetherington later recalled. 'Everyone laughed and soon the cat was out of the bag.'

Some years later, Hetherington and his collaborators were struggling to agree on the finer points of an article they were working on. With none of them ultimately willing to sign off on the finished product, they pulled Willard out of retirement and named him as the sole author of the paper, which was eventually published in the French journal *La Recherche*.[16]

Willard was considered for a position at the university and, in honour of his contribution to physics, APS Journals announced (on 1 April, 2014) that all feline-authored publications would be made open access.[17] The announcement reads: 'Not since Schrödinger has there been an opportunity like this for cats in physics.'

Cat research

When cats aren't contributing to academic life, they are themselves the subject of a large body of interesting research (including a much-publicised study suggesting that your cat may wish to kill you).[*18] Feline-themed papers include 'Demography and Movements of Free-Ranging Domestic Cats in Rural Illinois' and 'How Cats Lap: Water Uptake by *Felis catus*'.[19] However, the most pressing cat research from a human perspective investigates their propensity for spreading mind-controlling parasites.

Cats are carriers of the parasite *Toxoplasma gondii*, which alters the behaviour of animals to make them less afraid of predators (and therefore more likely to be killed, eaten, and used as a conduit for further propagation of the parasite).[†20] An unconventional Czech scientist, Jaroslav Flegr, has

---

* 'Cats ARE neurotic – and they're probably also trying to work out how to kill you, say researchers' (I am concerned that citing the *Daily Mail* twice in one ostensibly academic book is going to cause a rift in the time-space continuum). In fact, the study in question simply says that domestic cats share personality traits with lions, but viral clickbait the truth does not make.

† Scientists recently discovered a similar mechanism used by the pathogenic fungus *Batrachochytrium dendrobatidis*, in its infection of amphibians. Researchers found that Japanese tree frogs infected by the fungus exerted greater effort in their mating calls, and that their calls were faster and longer (which the female frogs prefer). This means that infected frogs tend to attract more females and therefore reproduce quicker, further spreading the fungus.

made researching these parasites his life's work. Ever since a light bulb moment in the early 1990s, he has been investigating the potentially parasitic link between cats and humans.[21] We've long understood that infection with *Toxoplasma* is a danger during pregnancy and a major threat to people with weakened immunity. However, the research of Flegr and others goes further, suggesting that infected humans are statistically more likely to be involved in car crashes caused by dangerous driving and have greater susceptibility to schizophrenia and depression.[22]

Even if your cat is trying to kill you or is inadvertently depressing you, they are still cute, and looking at cute pictures has been shown to improve your productivity.[23] Kitty pics will always leave you feline good.*

PLAYING FOWL

Chickens prefer beautiful humans. That is the conclusion (and title) of a 2002 paper published in *Human Nature*.[24] The researchers trained chickens to identify humans by pecking at a photo of an average face on a computer screen in exchange for food. Then, when the chickens were presented with a mix of photos, they pecked more at the photos of attractive faces (as determined by asking a group of biology undergraduates which people they would like to go on a date with). The import of the study, which is not immediately obvious, is that 'Human preferences arise from general properties of nervous systems, rather than from face-specific adaptations.' In a similar fashion, pigeons can be taught to discriminate between good and bad paintings by children.[25]

If all this seems rather odd, consider the presentation given by Doug Zongker during the humour session at the 2007 conference of the American Association for the Advancement of Science. Zongker's presentation consists entirely of the word 'chicken' repeated over and over, as do his slides, which also feature nonsensical chicken flow charts and graphs.[26] At the end of his presentation an audience member asks if the research was funded by

---

* #SorryNotSorry.

Colonel Sanders, to which Zongker replies: 'Chicken.'[*27]

In her thesis on 'Evaluating Computational Creativity',[28] Anna Jordanous uses Zongker's paper as an example of how humour differs across domains: '*Chicken* shows creativity in a domain that emphasises content correctness and usefulness (scientific research papers), because of the extreme absence of any scientifically useful and correct content.' She isn't the only one to have cited Zongker's epizeuxical paper. Evan Bradley slipped a reference to *Chicken* into his PhD thesis,[29] and now includes it in the reading lists for his psychology classes at Penn State Brandywine.[30] In *A Field Guide to Mesozoic Birds and Other Winged Dinosaurs*, the authors

---

* Zongker had previously published 'Chicken Chicken: Chicken Chicken Chicken' as a paper in the *Annals of Improbable Research*. I believe the paper should have been rejected at the peer review stage, as it does not mention relevant previous work conducted in Dmitri Borgmann's *Beyond Language: Adventures in Word and Thought* (1967). In *Beyond Language*, Borgmann notes that 'Buffalo buffalo Buffalo buffalo buffalo buffalo Buffalo buffalo', is grammatically correct in English, using it to demonstrate how homonyms and homophones can be used to create complicated linguistic constructs. The sentence plays on three possible meanings of the word buffalo: the animal; the city in New York; and the rather uncommon verb, to buffalo (i.e. to bully or intimidate, or to baffle). The sentence uses a restrictive clause (thus there are no commas and the word 'which' is omitted (e.g. 'Buffalo buffalo, *which* Buffalo buffalo buffalo')) and is also a reduced relative clause (i.e. the word 'that', which could appear between the second and third words of the sentence, is omitted). The sentence says that buffalo that are bullied by other buffalo are themselves bullying buffalo (in the city of Buffalo). In other words, *the* buffalo *from* Buffalo *which are* buffalo*ed by* buffalo *from* Buffalo, buffalo (verb) *other* buffalo *from* Buffalo. Tymoczko et al's 1995 book *Sweet Reason: A Field Guide to Modern Logic* argues that there is nothing significant about eight buffalo, as any sentence consisting solely of the word 'buffalo' repeated any number of times is grammatically correct, such that 'Buffalo buffalo Buffalo buffalo buffalo buffalo Buffalo buffalo Buffalo buffalo Buffalo buffalo buffalo buffalo Buffalo buffalo Buffalo buffalo Buffalo buffalo buffalo buffalo Buffalo buffalo Buffalo buffalo Buffalo buffalo buffalo buffalo Buffalo buffalo Buffalo buffalo Buffalo buffalo buffalo buffalo buffalo buffalo Buffalo buffalo Buffalo buffalo Buffalo buffalo buffalo buffalo Buffalo buffalo Buffalo buffalo Buffalo buffalo buffalo buffalo Buffalo buffalo Buffalo buffalo Buffalo buffalo buffalo buffalo buffalo buffalo Buffalo buffalo Buffalo buffalo Buffalo buffalo buffalo buffalo Buffalo' is grammatically correct, if a little gratuitous. The shortest possible sentence is 'Buffalo!', an imperative instruction to bully someone.

cite *Chicken* as their source for the statement: 'Even the yellow yolk of a chicken egg is due to carotenoids.'[31]

(HOMOSEXUAL NECROPHILIAC) DUCKS

I've always loved ducks (they can fly and their body is a boat – what's not to love?), but a notorious study has tested this love.

One June day in 1995, at around 5.55p.m., Cees Moeliker was happily working away at the Natuurmuseum Rotterdam when he heard an almighty thud. These noises were not uncommon. The genius architects that designed the new wing of the museum, situated in the middle of a park, had decided that it would look great in glass. Unfortunately, when the sun is shining the glass acts as a mirror, so birds don't see it and sometimes collide head on.

Moeliker went to check the situation and spotted a dead duck. He describes the next moments:

> *Next to the obviously dead duck, another male mallard (in full adult plumage without any visible traces of moult) was present. He forcibly picked into the back, the base of the bill and mostly into the back of the head of the dead mallard for about two minutes, then mounted the corpse and started to copulate, with great force, almost continuously picking the side of the head.*

Moeliker then did what any good researcher would do:

> *Rather startled, I watched this scene from close quarters behind the window until 19.10 during which time (75 minutes!) I made some photographs.*

He noted that the duck dismounted only twice during this time, resting for a matter of minutes before recommencing. A search of the literature revealed that, while ducks both engage in homosexual and necrophilic activities, nobody had ever documented a case of homosexual necrophilia in the mallard.

Moeliker's paper won him an Ig Nobel Prize in 2003 and he's given a TED talk about his experience.[32] Composer Dan Gillingwater wrote a mini-opera based on the incident that explores sexual attraction in the natural world,[*][33] and the museum now holds an annual Dead Duck Day.[34]

RATS

Lab rats and mice are the workhorses of science, being subjected to all sorts of horrible and nonsensical acts in our pursuit of knowledge. This being a supposedly humorous book, I do not wish to dwell on the fates of the millions of animals used in labs each year. What I do want to dwell on is the fact that sometimes, when they are not being genetically modified or running around mazes, lab rats are hanging out in tiny trousers, or are being tickled, for science.

Back on page 4, you saw the most glorious figure ever to grace the pages of a scientific journal, 'The underpant worn by the rat'. The keywords for the paper containing this incredible diagram include 'penis', 'erection' and 'electrostatic potentials', but fail to mention rats in underpants. The paper title, 'Effects of Different Types of Textiles on Sexual Activity', is more revealing.[35] Previous research on humans suggested that the electrostaticity generated by polyester underwear could render a man's sperm useless in five months, and Ahmed Shafik of Cairo University decided to investigate.

Shafik's illustration first gained international infamy after Mary Roach discussed his research in *Bonk: The Curious Coupling of Sex and Science*.[36] Roach highlighted Shafik's 'strange, brave career' noting that he had published over a thousand papers on such a diverse range of topics that it

---

* '… in an avant-garde pseudo-operatic, quasi-musical theatre/soul/funk style, this is a musical experience not to be missed! This event is for those aged over 18.'

is impossible to pin down his speciality.*

Shafik rounded up a group of 75 male rats, some of whom† wore tiny polyester underpants. (If your imagination is overactive like mine, this is the point at which you are picturing miniature washing machines and tiny wardrobes …) The other rats wore pants of cotton, wool, or a 50/50 polyester cotton mix. One lucky group evaded underpants altogether. At 6- and 12-month periods the rats were introduced to lady rats and their behaviour was recorded. The rats in the polyester and mixed pants were definitely feeling the love and were quick to mount their mates, but they finished the job much less often than their cotton-panted counterparts.

Shafik reckoned that the rodents woes were caused by static electricity building up in the pants, but concerned scientists on the internet doubt the veracity of this claim.[37] Various alternative explanations were offered, including the deleterious effects of an increase in heat, and the embarrassment of having to wear the pants.

We don't know for sure if rats feel embarrassment, but one team of neuroscientists has been trying to figure out if they might feel happiness. In the late 1990s neuroscientist Jaak Panksepp and his colleagues were thinking about how human emotions can cause subconscious biases in our thinking and decision-making and wondered if animals might show similar biases.[38] That is difficult to test because we don't have any way to ask a rat whether it is as happy as Larry or as down in the dumps as a farmed salmon.‡ But now we do: rats laugh.

---

* Roach writes: 'If you ask him what he is, what he writes under "Occupation" on his tax form, he will smile broadly and exclaim, "I am Ahmed Shafik!"'

† My word processor tells me that this should be 'which'. I understand that the grammatical convention is to use 'which' where the subject is non-human or not a pet, but I feel that when rats start wearing trousers they are sufficiently anthropomorphised to justify this small linguistic shift. Indeed, having been forced against their will to wear polyester underpants for extended periods, I feel it only fair to restore a shred of their dignity through more generous linguistic conventions.

‡ See footnote on page 122. If you haven't been reading the footnotes, you have been missing out.

Or at least we think they do. Panksepp found that rats emit 50kHz ultrasonic 'chirps' while playing with other rats, or even when they are anticipating playing with other rats. They also chirp when they are subjected to 'playful, experimenter-administered, manual, somatosensory stimulation' (i.e. tickling). In fact, rats laugh more when being tickled by people than when they are playing with other rats.

> *The tickling was done with the right hand and consisted of rapid initial finger movements across the back with a focus on the neck, followed by rapidly turning the animals over on their backs, with vigorous tickling of their ventral surface, followed by release after a few seconds of stimulation. This was repeated throughout each tickling session. Even though the tickling was brisk and assertive, care was taken not to frighten the animals.*

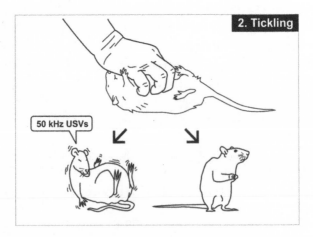

Figure 18: Playful, experimenter-administered, manual, somatosensory stimulation of *Rattus norvegicus*

In an attempt to discern the meaning of this laughter, the researchers trained rats to press a lever in response to a tone in order to obtain some food, and to press a second lever in response to a different tone to avoid an unpleasant electric shock to the foot.* Once the rats reliably knew the difference, they were divided into two groups, tickled and non-tickled, and presented with an ambiguous tone. The rats that laughed a lot when tickled were more optimistic, generally assuming they would be fed when the tone was ambiguous.

Two other interesting insights come from this study. Firstly, some of the tickled rats didn't seem to like being tickled and didn't respond with laughter. Secondly, whether or not a rat laughs when tickled is a stable behavioural trait that can be selected for. That means that in just four generations, we can breed rats that love to be tickled. These tickleable rats tend to play more, laugh more, and can learn faster when tickling is the reward.

PENGUINS

There is a huge body of research on penguins that can keep you (or at least me) amused for hours. Recent finds include decoding of a 'language' used by jackass penguins and discovery of fossils of a giant two-metre tall penguin.[39]

Some penguins, in particular chinstrap and Adélie penguins, appear to defecate fairly forcefully, a fact that proved worthy of further study to Victor Benno Meyer-Rochow and Jozsef Gal, who published a dedicated paper on the matter in *Polar Biology*.[40] Meyer-Rochow describes how the paper, 'Pressures Produced when Penguins Pooh: Calculations on Avian Defaecation', came about:[41]

> *Our project started in Antarctica during the first (and only) Jamaican Antarctic Expedition in 1993 . . . Many photographs of penguins and their 'decorated' nests were taken. Later at a slide show . . . I was asked by a student*

---

* Poor little guys :(

> *during question time to explain how the penguins decorated*
> *their nests. I answered: 'They get up, move to the edge of the*
> *nest, turn around, bend over – and shoot…' She blushed,*
> *the audience chuckled, and we got the idea to calculate the*
> *pressures produced when penguins poo.*

As with many humorous papers, *'Pressures Produced when Penguins Pooh'* elicited a number of genuine scientific research questions and follow-ups. A palaeontologist studying dinosaur biology thought that the calculations could be applied to similar streaks found near fossil dinosaur nests, zoo-operators enquired about safe distances for visitors, and a medical researcher was inspired to recalculate the same measures for humans (it had been done previously, but the data was quite old).

Penguin poo also turned out to have another useful purpose: locating penguin colonies from space. In a paper entitled 'Penguins from Space: Faecal Stains Reveal the Location of Emperor Penguin Colonies', researchers used satellite imagery to spot the distinctive brown stains left by emperor penguin colonies. Using this technique, they were able to better understand the position of six known locations, as well as rule out six old locations and identify ten new colonies.

These beautiful birds have also waddled their way into some obscure corners of academia. One innovative use of the penguin's likeness comes from particle physics, where it is used to represent weak decay of particles. Originally, these diagrams looked nothing like penguins, but that changed when John Ellis, now a professor of theoretical physics at King's College London, went for a drink with Melissa Franklin and Serge Rudaz.

As he recalls:[42]

> *Melissa and I started a game of darts. We made a bet that if*
> *I lost I had to put the word penguin into my next paper. She*
> *actually left the darts game before the end, and was replaced*
> *by Serge, who beat me. Nevertheless, I felt obligated to carry*
> *out the conditions of the bet.*

Rudaz later recounted that for him to beat Ellis at a game of darts was nothing short of miraculous: John was a strong player and even brought his own set of darts to the pub. His surprise victory meant that Ellis had to find a way to work penguins into his next paper.

> *For some time, it was not clear to me how to get the word into this b quark paper that we were writing at the time. Then, one evening, after working at CERN, I stopped on my way back to my apartment to visit some friends living in Meyrin where I smoked some illegal substance. Later, when I got back to my apartment and continued working on our paper, I had a sudden flash that the famous diagrams look like penguins. So we put the name into our paper, and the rest, as they say, is history.*

Not to be outdone by physicists, chemists got in on the joke. Having realised that 3,4,4,5-tetramethylcyclohexa-2,5-dien-1-one was a dull name for a chemical and that its 2-dimensional molecular structure resembled a penguin, they gave it the common name *penguinone*.

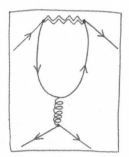 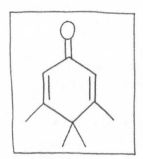

Figure 19: Feynman diagram of bottom quark decay and 2-dimensional formula of 3,4,4,5-tetramethylcyclohexa-2,5-dien-1-one

OBSCURE INTERLUDE

MISCELLANY

Skateboarding profs: Thomas Winter, a 68-year-old associate professor of classics and religious studies at the University of Nebraska-Lincoln, got his 15 minutes of internet fame when a photo of him riding his skateboard around campus was posted to Reddit.[1] The post garnered over 1,000 comments and the photo became a meme of its own, with people pairing his photo with skateboarding puns ('No test today, board meeting'; 'Summer's over, back to the grind').[2] His reviews on Rate My Professors are full of comments like 'Insane but intelligent' and 'Seriously crazy, but a lot of fun', as well as testaments to his teaching. Not slowing down in retirement, Winter is spending his time skydiving, welding, and flying his small plane around the US.[3]

Student living: University College London's New Hall housing complex won the Carbuncle Cup for the UK's worst building in 2013.[4] The hulking great £18-million building was originally refused planning permission for a long list of reasons, including its excessive scale, inadequate daylight, poor outlook and lack of privacy (the bedrooms face their neighbours' windows as close as five metres away – in a borough where the minimal residential overlooking distance is 18 metres). It could be worse. Goce Delcev student dormitory in Skopje, Macedonia, is the largest in the country, housing over 1,200 students every year. Photos posted online show the place damp and decaying, the walls peeling and turning green from mould, with lights and radiators that often don't work.[5] At the entrance there are two signs: 'There is no warm water. The problem is being fixed' and 'Go take a shower at your boyfriends' places!'

Thunderbirds are Go!: *Rapid Interpretation of EKG's* is a best-selling textbook that teaches fledgling doctors the basics of interpreting electrocardiograms. In its 50th printing, the author, millionaire plastic surgeon Dale Dublin,[*][6] included a picture of his 1965 Ford Thunderbird to explain electrode placement. He also hid a message in the fine print, promising to enter anybody that found it into a competition to win the car. Even though 60,000 copies were sold, less than half a dozen people wrote in. Yale medical student Jeffrey Seiden was drawn at random, presented the prize by Dublin's daughter, and rode off into the sunset while blasting the Beach Boys over the stereo.[7]

Staplers: Jason Vance, librarian and assistant professor at Middle Tennessee State University, started a blog entitled 'The Lives and Deaths of Academic Staplers',[8] in which he tracks the public staplers at the university library. Highlight: 'One notable holdover from the Spring 2015 semester study is Swingline's "Optima 70". It is currently 484 days old and is still going strong at the reference desk.'

Working with academics: During a presentation at a conference for start-ups, Aim Sinpeng, a political scientist at the University of Sydney, snapped a photo of the following slide:

---

[*]   Dublin was also a champion hibiscus grower and a felon convicted of 22 counts of drug and child pornography charges.

Working with Academics

- Annoying

- Terrible time management

- Have different goals

- Usually don't have money

CONCLUSION

I've never much cared for conclusions. I don't like to read them, much less write them, and it's not easy to elegantly transition from penguin defecation and wayward staplers to a neat closing statement. I tried the usual trick of pulling together a summary of highlights from the body of the text, but it doesn't seem fitting here:

> *In chapter two I ranted about clichés, while in chapter three we laughed at a terrible journal created by a suspended student. In chapter four I wrote about not writing and swore profusely… Elsewhere, rats wore pants and were tickled, for science, and cats wore parachutes because statistics is boring …*

If I am supposed to claim to have contributed to the sum of human knowledge in some profound way, or to have proposed a grand new theory of life, the universe and everything, I fear that all I have to offer is this: You don't have to be mad to work here, but you probably are.*

I implore you to own this insanity. Send a silly academic tweet, study an improbable topic, or include a humorous reference in your next paper.

---

* And it probably doesn't help.

Lightening up has the brilliant benefit of making academia interesting and accessible, and you don't necessarily even need to be that witty or clever to capture people's imagination.

If you aren't already rushing to the office to immediately do all of the above, I hope that I have at least inspired you to embrace humour in your work and to take academia a little less seriously. If that means you get a brief break from the daily grind, feel a bit more creative, and ultimately add to my growing folder of amusing academic obscurities that I fondly flick through from time to time, I'll consider this a success.

Right then. Back to the PhD.

IRRELEVANT BIBLIOGRAPHY

- Janet and Allan Ahlberg, *The Ha Ha Bonk Book* (1982)
- Linda Benedik, *Yoga for Equestrians: A New Path for Achieving Union with the Horse* (2000)
- Don Colbert MD, *The Bible Cure for Irritable Bowel Syndrome: Ancient Truths, Natural Remedies and the Latest Findings for Your Health Today* (2002)
- Jack Douglas, *The Jewish-Japanese Sex and Cook Book and How to Raise Wolves* (1974)
- Francis Johnson, *Historic Staircases in Durham City* (1970)
- Graham Johnson and Rob Hibbert, *Images You Should Not Masturbate To* (2011)
- Hiroyuki Nishigaki, *How to Good-bye Depression: If You Constrict Anus 100 Times Everyday. Malarkey? or Effective Way?* (2000)
- Michael Rosen, *Don't Put Mustard in the Custard* (1996)
- Redcliffe Salaman, *The History and Social Influence of the Potato* (1949)
- John Trimmer, *How to Avoid Huge Ships* (1982)
- Roger Welsch, *Everything I Know About Women I Learned from my Tractor* (2002)

PEER REVIEW REPORT

Reviewer 1

Overall, this is a decent contribution to the literature on academic humour, with some interesting and unusual stories. I even chuckled aloud on occasion. I nonetheless feel that the author has neglected to include some key references and could make some changes to improve the manuscript:

- Author does not cite Pfaus & Zunino (2014), who substantially advance Shafik (1993) by dressing rats in lingerie.[1]

- Consider including a few lines discussing chemicals and minerals with funny names,[2] e.g. Arsole (which, I am led to believe, is lightly aromatic),[3] Moronic Acid, Spamol, Cummingtonite, Fukalite, Diabolic Acid, Welshite.

- Far too many pointless footnotes. Delete at least half of them.

- Section on journal stings should include mention of Peters & Ceci (1982).[4] The authors resubmitted twelve published papers from prestigious US psychology schools to highly regarded journals. They used false names and institutions, and resubmitted them to the same journal that had reviewed and published them 18 to 32 months earlier. Of 38 editors/ reviewers, only three (8%) noticed they were resubmissions. Of the nine papers allowed to continue to peer review, eight were rejected. In many cases, the grounds for rejection were 'serious methodological flaws'.

- The author will no doubt be sad to learn that the whereabouts of Jordan the Edinburgh University library cat are currently unknown.[5]

- The author will however be amused to learn that Southampton University Students' Union has recognised a cat as its Honorary President.[6]

- On the subject of cats, the assertion that they cause mental illness is absolute nonsense. The author is clearly not familiar with a recent cohort study that found no evidence of an association between cat ownership and psychotic symptoms.[7]

- The author includes lengthy discussion of Beall's list, apparently unaware that it is now defunct.[8]

- Perhaps not appropriate for inclusion in a humorous publication, but given the recurring theme of institutionalised sexism, the author may wish to mention Clancy et al. (2014).[9] Researchers surveyed over 650 field scientists, finding that 71% of the female respondents had experienced harassment at field sites and 36% had been physically assaulted (the figure for male respondents was 41% and 6% respectively).[♀]

- The author could use footnotes more.

- John Mauchly, co-inventor of the first electronic digital computer (ENIAC), would skate across lab benches on a rocket-propelled skateboard to demonstrate principles of physics. As this was before the invention of the modern skateboard, he was technically one of the first to create one.[10]

- The author should include reference to the Journal of Alternative Facts and the latest paper published therein, 'We Have All the Best Climates, Really, They're Great'.[11]

- The author, like many an academic, jokes about the use of the Comic Sans. I once shared this misguided distaste for the much-maligned font, but recently learned that it can in fact be very useful for people with dyslexia.[12]

- I am surprised to see no mention of the 'Dr Fox Effect', so named after one of the first studies into the effect of lecturer charisma on student evaluations.[13] Researchers coached an actor to give a lecture on an irrelevant topic ('Mathematical Game Theory as Applied to Physician Education') to a class of psychiatrists and psychologists. Under the Dr Fox pseudonym, the actor gave an empty lecture 'with an excessive use of double talk, neologisms, non sequiturs, and contradictory statements',[14] yet the students submitted teaching evaluations that were overwhelmingly positive.

- I am equally surprised to see no mention of Polly Matzinger. Matzinger added her Afghan Hound, Galadriel Mirkwood, as a co-author on a paper,[15] though this may not have been completely without merit: while working on her well-known 'danger model' of immunology she suddenly realised that dendritic cells behave in the same way as a sheepdog. Her tenure committee later saw the funny side and decided that 'it wasn't really fraud. It was a real dog, a frequent lab visitor, and they said it had done no less research than some other coauthors had'.[16] Polly remains an avid sheepdog trainer and along with her two Border Collies, Charlie and Lily, was part of the US team at the 2005 World Sheepdog Finals.

Decision: Accept with minor revisions.

Reviewer 2

- Author does not provide an explanation for the non-capitalisation of 'Internet'.
- Excessive use of footnotes.
- Too much focus on cats.

Decision: Reject.

ANNEX I:
SELECTED FIGURES

256: Freud's h-index

1.5–8.7psi: Pressure produced during chinstrap penguin defecation

1,525: Number of papers authored by Paul Erdős.

2800+: Number of citations to the leading textbook on design of pavements, Huang's *Pavement analysis and design* (1993).

A metric fuckton: Profits made by academic publishers each year.

25–65%: The percentage range of: army recruits sustaining musculoskeletal injury during basic training; the range of exploitation rates of walleyes in Henderson and Savanne Lakes (Thunder Bay, Ontario); complication rates in skull base surgery and reconstruction; and reduction in pesticide use on onions when integrated pest management is implemented.[1]

4%: Number of funding applications rejected by the UK's Natural Environment Research Council each year due to the applicant using the incorrect font and formatting in their application.[2]

3.6 million: Number of hours I estimate that I wasted going on unrelated tangents during the writing of this book.

4: Number of papers written by US President Barack Obama while he was in office between 2009–2017.[3]

£26,000: Average undergraduate tuition fees for a degree in the UK since the cap was raised to £9,000 per year in 2012.[4]

ANNEX II: BUCKET LIST

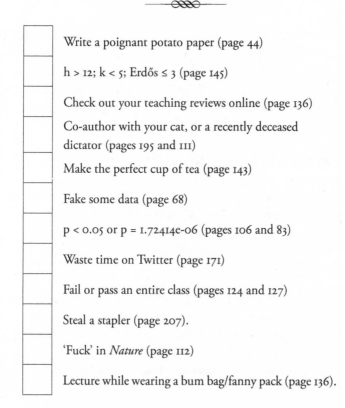

☐	Write a poignant potato paper (page 44)
☐	$h > 12$; $k < 5$; Erdős ≤ 3 (page 145)
☐	Check out your teaching reviews online (page 136)
☐	Co-author with your cat, or a recently deceased dictator (pages 195 and 111)
☐	Make the perfect cup of tea (page 143)
☐	Fake some data (page 68)
☐	$p < 0.05$ or $p = 1.72414e-06$ (pages 106 and 83)
☐	Waste time on Twitter (page 171)
☐	Fail or pass an entire class (pages 124 and 127)
☐	Steal a stapler (page 207).
☐	'Fuck' in *Nature* (page 112)
☐	Lecture while wearing a bum bag/fanny pack (page 136).

NOTES

For the love of trees, I have opted to keep this bibliography (relatively) short. For more details, please go to AcademiaObscura.com/buffalo, where I plan to concoct a multimedia extravaganza containing links, photos, and videos. If I get distracted and don't get around to doing this (highly likely), I will at the very least provide full references and PDFs (where I can do so legally).

I.   WHAT'S ALL THIS NONSENSE THEN?

1   Alvesson and Spicer, 'A Stupidity-Based Theory of Organizations' (2012) *Journal of Management Studies*.

2   Davies & Blackwell, 'Energy Saving through Trail Following in a Marine Snail' (2007) *Proceedings of the Royal Society B: Biological Sciences*.

3   McConnell, *Science, Sex, and Sacred Cows: Spoofs on Science from the Worm Runner's Digest* (1971).

4   See, e.g. Schwartz, '"Sonic Hedgehog" Sounded Funny, at First' (2006) *New York Times*; Heard, 'On Whimsy, Jokes, and Beauty: Can Scientific Writing Be Enjoyed?' (2014) *Ideas in Ecology and Evolution*; Riesch, 'Why Did the Proton Cross the Road? Humour and Science Communication' (2014) *Public Understanding of Science*.

5   Connor, 'French Scientist Admits to Making up Saucy Acronyms for Genetics Research Papers as Part of a Dare' (2014) *Independent*.

II.   PUBLISH OR PERISH

1   Wilson, *The Academic Man: A Study in the Sociology of a Profession* (1942). For discussion, see Plume & Weijen, 'Publish or Perish? The Rise of the Fractional Author' (2014) *Research Trends*.

2   Carskadon, 'Sleep in Adolescents: The Perfect Storm' (2011) *Pediatric Clinics of North America*; Ferre, 'Alcohol and Caffeine: The Perfect Storm' (2011) *Journal of Caffeine Research*.

3   Conway & Murphy, 'A Rising Tide Meets a Perfect Storm: New Accountabilities in Teaching and Teacher Education in Ireland' (2013) *Irish Educational Studies*; Keane et al., 'Leading a Sea Change in Naval Ship Design: Toward Collaborative Product Development' (2007) *Journal of Ship Production*; Smith, Dawn Bazely & Yan, 'Missing the Boat on Invasive Alien Species: A Review of Post-Secondary Curricula in Canada' (2011) *Canadian Journal of Higher Education*.

4   Atkin, 'A Paradigm Shift in the Medical Literature' (2002) *British Medical Journal*.

5   Goodman, 'Familiarity Breeds: Clichés in Article Titles' (2012) *British Journal of General Practice*.

6   Goodman, 'From Shakespeare to Star Trek and beyond: A Medline Search for Literary and Other Allusions in Biomedical Titles' (2005) *British Medical Journal*.

7   Wilson et al., 'Much Ado about the Null Hypothesis' (1967) *Psychological Bulletin*.

8   'On This Day, 22 February, 1997: Dolly the Sheep Is Cloned', *BBC*.

9   Rubin, 'To Test or "NOD-2" test: What Are the Questions? The Balanced Viewpoint' (2005) *Inflammatory Bowel Diseases*.

10   Beijerinck et al., 'Breast Cancer Screening: All's Well That Ends Well, or Much Ado about Nothing?' (1989) *American Journal of Roentgenology*.

11   Jones & Miller, 'The Lateral Ligaments of the Rectum: The Emperor's New Clothes?' (2001) *Diseases of the Colon and Rectum*; Kavanagh, 'The Emperor's New Isodose Curves' (2003) *Medical Physics*.

12   Morle, 'Mentorship – Is It a Case of the Emperors New Clothes or a Rose by Any Other Name?' (1990) *Nurse Education Today*.

13   Gambrill, 'Evidence-Based Practice: Sea Change or the Emperor's New Clothes?' (2003) *Journal of Social Work Education*.

14   Giraldo-Perez et al., 'Winter Is Coming: Hibernation Reverses the Outcome of Sperm Competition in a Fly' (2016) *Journal of Evolutionary Biology*.

15   Schneider, 'And the Winner is "The Good the Bad and the Outsourced"' (2010).

16   Brand-Miller et al., 'Carbohydrates – the Good, the Bad and the
     Wholegrain' (2008) *Asia Pacific Journal of Clinical Nutrition*; Vangeison
     et al., 'The Good, the Bad, and the Cell Type-Specific Roles of Hypoxia
     Inducible Factor-1 Alpha in Neurons and Astrocytes' (2008) *The Journal of
     Neuroscience: The Official Journal of the Society for Neuroscience*.

17   Eidsmoe & Edwards, 'Sex, Lies, and Insurance Coverage? Insurance Carrier
     Coverage Defenses for Sexually Transmitted Disease Claims' (1999) *Tort &
     Insurance Law Journal*.

18   Hetterscheid & Ittenbach, 'Everything You Always Wanted to Know
     about Amorphophallus, but Were Afraid to Stick Your Nose Into!' (1996)
     Aroideana.

19   Petretti & Prigent, 'The Protein Kinase Resource: Everything You Always
     Wanted to Know about Protein Kinases but Were Afraid to Ask' (2005)
     Biology of the Cell.

20   Amsen, 'Of Mice and Men – a Poem' (2015) *Easternblot.net*.

21   Berger, 'Of Mice and Men: An Introduction to Mouseology Or, Anal
     Eroticism and Disney' (1991) *Journal of Homosexuality*; Jones et al., 'Of
     Mice and Men: The Evolving Phenotype of Aromatase Deficiency' (2006)
     Trends in Endocrinology and Metabolism.

22   Adler & Cole, 'Designed for Learning: A Tale of Two Auto Plants' (1993)
     Sloan Management Review; Gilbert & Ivancevich, 'Valuing Diversity: A Tale
     of Two Organizations' (2000) *Academy of Management Perspectives*.

23   Wang & Reynolds, 'Avoiding the "Catch 22" in Special Education Reform'
     (1985) *Exceptional Children*; Laimoo, 'Amphibian Conservation and
     Wetland Management in the Upper Midwest: A Catch-22 for the Cricket
     Frog' (1998) *Status and Conservation of Midwestern Amphibians*.

24   Lundberg et al., 'Nitric Oxide and Inflammation: The Answer Is Blowing in
     the Wind' (1997) *Nature Medicine*.

25   Mooradian & Olver, '"I Can't Get No Satisfaction:" The Impact of
     Personality and Emotion on Postpurchase Processes' (1997) *Psychology &
     Marketing*.

26   Cleary et al., 'Editorial: Money, Money, Money: Not so Funny in the
     Research World' (2015) *Journal of Clinical Nursing*.

27   Holland et al., 'Smells Like Clean Spirit: Nonconscious Effects of Scent on
     Cognition and' (2005) *Psychological Science*.

28 Breen, 'A Stairway to Heaven or a Highway to Hell?: Heavy Metal Rock Music in the 1990s' (1991) *Cultural Studies*.

29 Vincent & Lailvaux, 'Female Morphology, Web Design, and the Potential for Multiple Mating in Nephila Clavipes: Do Fat-Bottomed Girls Make the Spider World Go Round?' (2006) *Biological Journal of the Linnean Society*.

30 Utzinger et al., 'An In-Depth Analysis of a Piece of Shit: Distribution of Schistosoma Mansoni and Hookworm Eggs in Human Stool' (2012) *PLOS Neglected Tropical Diseases*; Campos-Arceiz, 'Shit Happens (to Be Useful)! Use of Elephant Dung as Habitat by Amphibians' (2009) *Biotropica*.

31 Shalvi et al., 'Write When Hot – Submit When Not: Seasonal Bias in Peer Review or Acceptance?' (2010) *Learned Publishing*.

32 Hartley, 'Write When You Can and Submit When You Are Ready!' (2011) *Learned Publishing*.

33 Calabresi & Melamed, 'Property Rules, Liability Rules, and Inalienability: One View of the Cathedral' (1972) *Harvard Law Review*.

34 Krier & Schwab, 'Property Rules and Liability Rules: The Cathedral in Another Light' (1995) *NYUL Review*; Caruso, 'The Missing View of the Cathedral: The Private Law Paradigm of European Legal Integration' (1997) *European Law Journal*; Epstein, 'A Clear View of the Cathedral: The Dominance of Property Rules' (1997) *Yale Law Journal*; Nance, 'Guidance Rules and Enforcement Rules: A Better View of the Cathedral' (1997) *Virginia Law Review*; Rose, 'The Shadow of the Cathedral' (1997) *Yale Law Journal*; Schroedert, 'Three's a Crowd: A Feminist Critique of Calabresi and Melamed's One View of the Cathedral' (1999) *Cornell Law Review*; Bebchuk, 'Property Rights and Liability Rules: The Ex Ante View of the Cathedral' (2001) *Michigan Law Review*; Rule, 'A Downwind View of the Cathedral: Using Rule Four to Allocate Wind Rights' (2009) *San Diego Law Review*; Torrance & Tomlinson, 'Property Rules, Liability Rules, and Patents: One Experimental View of the Cathedral' (2011) *Yale Journal of Law and Technology*.

35 Farber, 'Another View of the Quagmire: Unconstitutional Conditions and Contract Theory' (2005) *Florida State University Law Review*.

36 Rayner et al., 'Raeding Wrods with Jubmled Lettres: There Is a Cost' (2006) *Psychological Science*.

37   Storms et al., 'Not Guppies, nor Goldfish, but Tumble Dryers, Noriega, Jesse Jackson, Panties, Car Crashes, Bird Books, and Stevie Wonder' (1998) *Memory & Cognition*.

38   Nazir & Chowdhary, 'From Urethra with Shove: Bladder Foreign Bodies. A Case Report and Review' (2006) *JAGS*.

39   Carlson et al., 'You Probably Think This Paper's about You: Narcissists' Perceptions of Their Personality and Reputation' (2011) *Journal of Personality and Social Psychology*.

40   Vale, 'Local Pancake Defeats Axis of Evil' (2005) *arXiv*.

41   THE Reporters, 'What's in an (Academic's) Name?' (2015) *Times Higher Education*.

42   Buttery et al., 'Studies on Popcorn Aroma and Flavor Volatiles' (1997) *Journal of Agricultural and Food Chemistry*; Cheeseman et al., 'Multiple Recent Horizontal Transfers of a Large Genomic Region in Cheese Making Fungi' (2014) *Nature Communications*.

43   Perris et al., 'Perceived Depriving Parental Rearing and Depression' (1986) *The British Journal of Psychiatry*; Abrahams, 'Things That Pop up in Databases. Read at Your Own Risk' (2013) *Improbable Research*.

44   Moran et al., 'Diffusion in a Periodic Lorentz Gas' (1987) *Journal of Statistical Physics*.

45   Geim & H.A.M.S ter Tisha, 'Detection of Earth Rotation with a Diamagnetically Levitating Gyroscope' (2001) *Physica B: Condensed Matter*. For a fascinating insight into Geim's work, see Lewis, 'Nobel Prize in Physics: Andre Geim Went from Levitating Frogs to Science's Highest Honor' (2014) *Slate*.

46   Brar et al., 'Observation of Carrier-Density-Dependent Many-Body Effects in Graphene via Tunneling Spectroscopy' (2010) *Physical Review Letters*; Jang et al., 'Tunable Large Resonant Absorption in a Midinfrared Graphene Salisbury Screen' (2014) *Physical Review B*.

47   'Name of the Year: Your 2011 Name of the Year' (2012). For Dutch names, see the Meertens Instituut Dutch Forename Database.

48   Manuwal et al., 'Progressive Territory Establishment of Four Species of Neotropical Migrants in Linear Riparian Areas in Western Montana' (2014) *BioOne*.

49   'A Family Affair: Four Manuwals Co-Author Paper' (2014) *Offshoots*.

50   Goodman et al., 'A Few Goodmen: Surname-Sharing Economist Coauthors' (2014) *Economic Inquiry*.

51   Scott et al., 'The Morphology of Steve' (2004) *Annals of Improbable Research*.

52   de Solla-Price, *Little Science, Big Science* (1963). Cronin, 'Hyperauthorship: A Postmodern Perversion or Evidence of a Structural Shift in Scholarly Communication Practices?' (2001) *Journal of the American Society for Information Science and Technology*.

53   Leung et al., 'Drosophila Muller F Elements Maintain a Distinct Set of Genomic Properties Over 40 Million Years of Evolution' (2015) *G3: Genes, Genomes, Genetics*.

54   Klionsky et al., 'Guidelines for the Use and Interpretation of Assays for Monitoring Autophagy (3rd Edition)' (2016) *Autophagy*.

55   ATLAS Collaboration & Collaboration, 'Combined Measurement of the Higgs Boson Mass in Pp Collisions at $s\sqrt{}=7$ and 8 TeV with the ATLAS and CMS Experiments' (2015) *Physical Review Letters*.

56   Castelvecchi, 'Physics Paper Sets Record with More than 5,000 Authors' (2015) *Nature News*.

57   Howe et al., 'Corrigendum: The Zebrafish Reference Genome Sequence and Its Relationship to the Human Genome' (2013) *Nature*.

58   Elmendorf, 'Corrigendum to Elmendorf et al. (2012)' (2014) *Ecology Letters*.

59   Gamow, *The Creation of the Universe* (1952).

60   'April 1, 1948: The Origin of Chemical Elements' (2008) *APS News*.

61   Singh, *Big Bang: The Origin of the Universe* (2005).

62   Singh, 'The Forgotten Father of the Big Bang' (2004) *Telegraph*.

63   Roderick & Gillespie, 'Speciation and Phylogeography of Hawaiian Terrestrial Arthropods' (1998) *Molecular Ecology*.

64   Hart, 'Co-Authorship in the Academic Library Literature: A Survey of Attitudes and Behaviors' (2000) *Journal of Academic Librarianship*.

65   Deville, 'How to Determine the Order of Authorship in an Academic Paper' (2014) *Sylvaindeville.net*.

66   Kersten & Earles, 'Semantic Context Influences Memory for Verbs More than Memory for Nouns' (2004) *Memory & Cognition*.

67   Baumeister et al., 'Subjective and Experiential Correlates of Guilt in Daily Life' (1995) *Personality and Social Psychology Bulletin*.

68   Swann et al., 'The Fleeting Gleam of Praise: Cognitive Processes Underlying Behavioral Reactions to Self-Relevant Feedback' (1990) *Journal of Personality and Social Psychology*.

69   Hassell & May, 'Aggregation of predators and insect parasites and its effect on stability' (1974) *Journal of Animal Ecology*.

70   Godfray, '100 Influential Papers – Longer Commentary. #13' (2012) *British Ecological Society*.

71   Schulman et al., 'Data Analysis Using S-Plus' (1995) *Sociological Methods & Research*.

72   Feder & Mitchell-Olds, 'Opinion: Evolutionary and Ecological Functional Genomics' (2003) *Nature Reviews Genetics*.

73   Holyoak & Walker, 'Subjective Magnitude Information in Semantic Orderings' (1976) *Journal of Verbal Learning and Verbal Behavior*.

74   Jolicoeur & Besner, 'Additivity and Interaction between Size Ratio and Response Category in the Comparison of Size-Discrepant Shapes' (1987) *Journal of Experimental Psychology: Human Perception and Performance*.

75   Griffiths & Anderson, 'Specification of Agricultural Supply Functions Empirical Evidence on Wheat in Southern N.S.W' (1978) *Australian Journal of Agricultural Economics*; Belyea & Lancaster, 'Inferring Landscape Dynamics of Bog Pools from Scaling Relationships and Spatial Patterns' (2002) *Journal of Ecology*; Kupfer et al., 'Forest Fragmentation Affects Early Successional Patterns on Shifting Cultivation Fields near Indian Church, Belize' (2004) *Agriculture, Ecosystems & Environment*.

76   Berry et al., 'Can Apparent Superluminal Neutrino Speeds Be Explained as a Quantum Weak Measurement?' (2011) *Journal of Physics A: Mathematical and Theoretical*.

77   Doyle, 'Guaranteed Margins for LQG Regulators' (1978) *IEEE Transactions on Automatic Control*.

78   Gardner & Knopoff, 'Is the Sequence of Earthquakes in Southern California, with Aftershocks Removed, Poissonian' (1974) *Bulletin of the Seismological Society of America*.

79   amarashiki, 'LOG#170. The Shortest Papers Ever: The List' (2015) *The Spectrum of Riemannium*.

80   Hajdukovic & Satz, 'Does the One-Dimensional Ising Model Show Intermittency?' (1992) *Nuclear Physics B*.

81 D'Amato et al., 'Harry Belafonte and the Secret Proteome of Coconut Milk' (2012) *Journal of Proteomics.*

82 Di Girolamo et al., 'Assessment of the Floral Origin of Honey via Proteomic Tools' (2012) *Journal of Proteomics.*

83 Email from Pier Righetti to Rajini Rao (21 March, 2014).

84 Faulkes, 'Maybe These Graphical Abstracts Could Be a Little Less Graphic' (2014) *NeuroDojo.*

85 Rodell, 'Goodbye to Law Reviews' (1936) *Virginia Law Review.*

86 Grafton, *The Footnote: A Curious History* (1999).

87 Brown, 'The Joy of Footnotes (1)' (2008) *Telegraph.*

88 Staley, *Reading with a Passion: Rhetoric, Autobiography, and the American West in the Gospel of John* (1995).

89 Louis & Sirico, 'Reining in Footnotes' (2005) *Perspectives*; Ames Magat, 'Bottomheavy: Legal Footnotes' (2010) *Journal of Legal Education.*

90 Pudwell, 'Digit Reversal Without Apology' (2005) *arXiv.*

91 Blanchard et al., *Differential Equations* (2011)

92 Yang et al., 'Law of Urination: All Mammals Empty Their Bladders Over the Same Duration' (2013) *arXiv.*

93 Yang et al., 'Duration of Urination Does Not Change with Body Size' (2014) *PNAS.*

94 Morisaka et al., 'Spontaneous Ejaculation in a Wild Indo-Pacific Bottlenose Dolphin (*Tursiops Aduncus*)' (2013) *PLOS ONE.* For spontaneous ejaculation in other animals, see: Aronson, 'Behavior Resembling Spontaneous Emissions in the Domestic Cat' (1949) *Journal of Comparative and Physiological Psychology*; Orbach, 'Spontaneous Ejaculation in a Rat' (1961) *Science*; Beach & Eaton, 'Androgenic Control of Spontaneous Seminal Emission in Hamsters' (1969) *Physiology & Behavior*; Huber & Bronson, *Social Modulation of Spontaneous Ejaculation in the Mouse* (1980) *Behavioral and Neural Biology*; McDonnel, 'Spontaneous Erection and Masturbation in Equids' (1989) in *Proceedings of the annual convention of the American Association of Equine Practitioners.* For the unusual human case, see Sivrioglu et al., 'Reboxetine Induced Erectile Dysfunction and Spontaneous Ejaculation during Defecation and Micturition' (2007) *Progress in Neuro-Psychopharmacology and Biological Psychiatry.*

95 Reisdorf et al., 'Float, Explode or Sink: Postmortem Fate of Lung-Breathing Marine Vertebrates' (2011) *Palaeobiodiversity and Palaeoenvironments*.

96 Maruthupandian & Marimuthu, 'Cunnilingus Apparently Increases Duration of Copulation in the Indian Flying Fox, *Pteropus giganteus*' (2013) *PLOS ONE*.

97 Utzinger et al., 'An In-Depth Analysis of a Piece of Shit: Distribution of Schistosoma Mansoni and Hookworm Eggs in Human Stool' (2012) *PLOS Neglected Tropical Diseases*.

98 Jablonski et al., 'Remains of Holocene Giant Pandas from Jiangdong Mountain (Yunnan, China) and Their Relevance to the Evolution of Quaternary Environments in South-Western China' (2012) *Historical Biology*.

99 Meyer-Rochow & Gal, 'Pressures Produced When Penguins Pooh – Calculations on Avian Defaecation' (2003) *Polar Biology*.

100 Shafik, 'Effect of Different Types of Textile Fabric on Spermatogenesis: An Experimental Study' (1993) *Urological Research*.

101 Due et al., 'Effect of Normal-Fat Diets, Either Medium or High in Protein, on Body Weight in Overweight Subjects: A Randomised 1-Year Trial' (2004) *International Journal of Obesity*.

102 Murakami-mizukami et al., 'Analyses of Indole Acetic Acid and Absclsic Acid Contents in Nodules of Soybean Plants Bearing VA Mycorrhizas' (1991) *Soil Science and Plant Nutrition*.

103 Xie et al., 'An Integrative Analysis of DNA Methylation and RNA-Seq Data for Human Heart, Kidney and Liver' (2011) *BMC Systems Biology*.

104 Monks et al., 'Variation in Melanism and Female Preference in Proximate but Ecologically Distinct Environments' (2014) *Ethology*.

105 'Overly Honest References: "Should We Cite the Crappy Gabor Paper Here?"' (2014) *Retraction Watch*.

106 Drinkel et al., 'Supporting Information for: Synthesis, Structure and Catalytic Studies of Palladium and Platinum Bissulfoxide Complexes' (2013).

107 Schulz, 'A Puzzle Named Bengü Sezen – A Historic Case of Fraud in the Chemistry Community Leaves Many Questions and Issues Unresolved' (2011) *ACS Publications*; Drahl et al., 'Insert Data Here ... But Make It Up First' (2013) *C&EN*.

108 Drahl, 'Controversial Organometallics Paper Cleared of Falsification Charge' (2014) *C&EN*; Gladysz & Liebeskind, 'Editors' Comments on the Addition/Correction to "Synthesis, Structure, and Catalytic Studies of Palladium and Platinum Bis-Sulfoxide Complexes"' (2014) *Organometallics*.

109 Drahl et al., 'Insert Data Here ... But Make It Up First' (2013) *C&EN*.

110 von Kieseritzky, 'In Defense of Emma' (2013) *Synthetic Remarks*.

OBSCURE INTERLUDE: ACADEMIC WHIMSY

1 Blackawton Primary School et al., 'Blackawton Bees' (2011) *Biology Letters*; Maloney & Hempel de Ibarra, 'Blackawton Bees: Commentary on Blackawton, P. S. et Al.' (2011) *Biology Letters*.

2 Shea et al., 'Pathology in the Hundred Acre Wood: A Neurodevelopmental Perspective on A.A. Milne' (2000) *Canadian Medical Association Journal*.

3 Miner, 'Body Ritual Among the Nacirema' (1956) *American Anthropologist*.

4 Krugman, 'The Theory of Interstellar Trade' (1978).

5 Feinstein, 'It's a Trap: Emperor Palpatine's Poison Pill' (2015).

6 Friedman & Hall, 'Using Star Wars' Supporting Characters to Teach about Psychopathology' (2015) *Australasian Psychiatry: Bulletin of Royal Australian and New Zealand College of Psychiatrists*.

7 Pope et al., 'Evolving Ideas of Male Body Images as Seen Through Action Toys' (1999) *International Journal of Eating Disorders*.

8 Griffiths et al., 'The Skywalker Twins Drift Apart' (2014) *Physics Special Topics*.

III. ACADEMIC PUBLISHING

1 Ware, *The STM Report: An Overview of Scientific and Scholarly Journal Publishing* (2015).

2 Jinha, 'Article 50 Million: An Estimate of the Number of Scholarly Articles in Existence' (2010) *Learned Publishing*.

3 Larivière et al., 'The Oligopoly of Academic Publishers in the Digital Era' (2015) *PLOS ONE*.

4 McGuigan & Russell, 'The Business of Academic Publishing' (2008) *Electronic Journal of Academic and Special Librarianship*.

5 Klein et al., 'Comparing Published Scientific Journal Articles to Their Pre-Print Versions' (2016) *arXiv*.

6   Smith, 'Librarians Find Common Ground with Former Foes' (2015)
     *Research.

7   Wright et al., 'This Study Is Intentionally Left Blank' (2015) *Annals of
     Improbable Research.*

8   Taylor, 'Elsevier Is Taking down Papers from Academia.edu' (2013) *Sauropod
     Vertebra Picture of the Week.*

9   Plumer, 'The World's Most Boring Journal – and Why It's Good for
     Science' (2012) *Washington Post.*

10   Flegel et al., 'Review of Disease Transmission Risks from Prawn Products
     Exported for Human Consumption' (2009) *Aquaculture*; Strom et al., 'The
     Female Menstrual Cycle Does Not Influence Testosterone Concentrations
     in Male Partners' (2012) *Journal of Negative Results in BioMedicine.*

11   Journal of Universal Rejection website.

12   Proceedings of the National Institute of Science website.

13   Department of Math & Theology (Matheology), 'Can You Pray Your Way
     towards Statistical Significance? An Experimental Test' (2015) *PNIS-HARD.*

14   Footman & Footman, 'Effects of Climate Change, Agricultural Clearing,
     and the Sun Becoming a Red Giant on an Old Growth Oak-Hickory Forest
     in Southeastern Iowa' (2014) *PNIS-SOFD.*

15   Tennant et al., 'The Academic, Economic and Societal Impacts of Open
     Access: An Evidence-Based Review' (2016) *F1000 Research.*

16   Xia et al., 'Who Publishes in "Predatory" Journals?' (2014) *Journal of the
     Association for Information and Science and Technology.*

17   'FTC Charges Academic Journal Publisher OMICS Group Deceived
     Researchers' (2016) *Federal Trade Commission*; McCook, 'Multiple OMICS
     Journals Delisted from Major Index over Concerns' (2017) *Retraction Watch.*

18   Schuman, 'Revise and Resubmit!' (2014) *Slate.*

19   Thanks to Twitter users Elana Halls (@hestofhere) and Jason Warr
     (@WarrCriminology) for the last two.

20   Bernstein, 'Updated: Sexist Peer Review Elicits Furious Twitter Response,
     PLOS Apology' (2015) *Science*; Woolston, 'Sexist Review Causes Twitter
     Storm' (2015) *Nature.*

21   Chapman & Slade, 'Rejection of Rejection: A Novel Approach to
     Overcoming Barriers to Publication' (2015) *British Medical Journal.*

22 'Gravitational Waves Detected 100 Years After Einstein's Prediction' (2016) *LIGO Lab website*.

23 Kennefick, 'Einstein Versus the Physical Review' (2005) *Physics Today*.

24 'Correction' (31 May 2016) *Nature*.

25 McCook, 'What Did Retractions Look like in the 17th Century?' (2016) *Retraction Watch*.

26 Oransky, 'The First-Ever English Language Retraction (1756)?' (2012) *Retraction Watch*.

27 'Müllsammler Der Wissenschaft' (2016) *Schweizer Radio Und Fernsehen*.

28 Oransky & Marcus, 'Why Write a Blog about Retractions?' (2010) *Retraction Watch*.

29 Schuman, 'Revise and Resubmit!' (2014) *Slate*.

30 Oransky & Marcus, 'AIDS Vaccine Fraudster Sentenced to Nearly 5 Years in Prison and to Pay Back $7 Million' (2015) *Retraction Watch*.

31 Oransky, 'Retraction of 19-Year-Old Nature Paper Reveals Hidden Cameras, Lab Break-In, Evidence Tampering – Retraction Watch at Retraction Watch' (2013).

32 Oransky, 'South Korean Plant Compound Researcher Faked Email Addresses so He Could Review His Own Studies' (2012) *Retraction Watch*.

33 Oransky, 'Retraction Count Grows to 35 for Scientist Who Faked Emails to Do His Own Peer Review' (2012) *Retraction Watch*.

34 Leung & Sharma, 'Education Minister Resigns over Research Fraud Scandal' (2014) *University World News*.

35 'You Can't Make this Stuff Up: Plagiarism Guideline Paper Retracted for . . . Plagiarism' (2015) *Retraction Watch*.

36 Shamim, 'Serious Thoughts about Plagiarism from India' (2012) *Saudi Journal of Anaesthesia*.

37 Marcus, 'Wikipedia Page Reincarnated as Paper: Authors Plagiarized Paper on Reincarnation' (2016) *Retraction Watch*; Rao, 'The Scourge of Rising Plagiarism' (2016) *The Hindu*.

38 Nagaraj et al., 'The Mystery of Reincarnation' (2013) *Indian Journal of Psychiatry*.

39 Borrell, 'A Bullshit Excuse? My Lab Notebook "was Blown into a Manure Pit"' (2016) *Retraction Watch*.

40   Butler, 'Mystery over Obesity "fraud"' (2013) *Nature*.

41   LaCour & Green, 'When Contact Changes Minds: An Experiment on Transmission of Support for Gay Equality' (2014) *Science*; McNutt, 'Editorial Retraction' (2015) *Science*.

42   'The Incredible Rarity of Changing Your Mind' (2015) *This American Life*.

43   Broockman et al., 'Irregularities in LaCour' (2014).

44   Shavin, 'Door-to-Door Deception' (2015) *New Republic*.

45   Leung & Sharma, 'Education Minister Resigns over Research Fraud Scandal' (2014) *University World News*.

46   Broockman & Kalla, 'Durably Reducing Transphobia: A Field Experiment on Door-to-Door Canvassing' (2016) *Science*.

47   Palus, 'BMC Retracts Paper by Scientist Who Banned Use of His Software by Immigrant-Friendly Countries' (2015) *Retraction Watch*.

48   Kupferschmidt, 'Scientist Says Researchers in Immigrant-Friendly Nations Can't Use His Software' (2015) *Science*.

49   Bhattacharjee, 'The Mind of a Con Man' (2013) *New York Times*.

50   Subramanian, 'Google Study Gets Employees to Stop Eating So Many M&Ms' (2013) *Time*.

51   Gattuso et al., 'Contrasting Futures for Ocean and Society from Different Anthropogenic CO_2 Emissions Scenarios' (2015) *Science*.

52   Ferguson, 'Article Using Tin Foil, Cling Wrap to Debunk Ocean Warming Retracted after Urgent Peer Review – Retraction Watch at Retraction Watch' (2014) *Retraction Watch*.

53   Fang et al., 'Misconduct Accounts for the Majority of Retracted Scientific Publications' (2012) *Proceedings of the National Academy of Sciences of the United States of America*.

54   Sokal, 'Transgressing the Boundaries: Towards a Transformative Hermeneutics of Quantum Gravity' (1999) *Social Text*.

55   Sokal, 'A Physicist Experiments With Cultural Studies' (1996) *Lingua Franca*.

56   Robbins & Ross, 'Response: Mystery Science Theater' (2000) in *The Sokal Hoax: The Sham that Shook the Academy*.

57   Sokal, 'A Physicist Experiments With Cultural Studies' (1996) *Lingua Franca*.

58 Abrahams, 'Words That, Taken Together Possibly Mean Something' (2014) *Improbable Research*.

59 Conner-Simons, 'How Three MIT Students Fooled the World of Scientific Journals' (2015) *MIT News*.

60 All of the emails and accompanying documentation are available on the SCIgen website.

61 Erickson, 'On Having a Huge Sadneess' (2005) *Ernie's 3D Pancakes*.

62 Simpson et al., '"Fuzzy" Homogeneous Configurations' (2014) *Aperito Journal of Nanoscience Technology*.

63 Seth & Singh, 'Use of Cloud-Computing and Social Media to Determine Box Office Performance' (2013).

64 Kabra, 'How I Published a Fake Paper, and Why It Is the Fault of Our Education System' (2013).

65 Labbé, SCIgen Detection website.

66 Sample, 'How Computer-Generated Fake Papers Are Flooding Academia' (2014) *Guardian*.

67 'Academic Publishing: Science's Sokal Moment' (2013) *Economist*.

68 Eisen, 'I Confess, I Wrote the Arsenic DNA Paper to Expose Flaws in Peer-Review at Subscription Based Journals' (2013) *Michaeleisen.org*.

69 Stromberg, '"Get Me Off Your Fucking Mailing List" is an Actual Science Paper Accepted by a Journal' (2014) *Vox*.

70 In *Sylvie and Bruno Concluded* (1893), illustrated by Harry Furniss. Public Domain: copy held at University of California Libraries (available at archive.org).

OBSCURE INTERLUDE: BEARDS

1 Janif et al., 'Negative Frequency-Dependent Preferences and Variation in Male Facial Hair Negative Frequency-Dependent Preferences and Variation in Male Facial Hair' (2014) *Biology Letters*.

2 Robinson, 'Fashions in Shaving and Trimming of the Beard: The Men of the Illustrated London News, 1842–1972' (1976) *American Journal of Sociology*.

3 Dowd, *Beards: An Archaeological and Historical Overview* (2010).

4 Dixson & Brooks, 'The Role of Facial Hair in Women's Perceptions of Men's Attractiveness, Health, Masculinity and Parenting Abilities' (2013) *Evolution and Human Behavior*.

5   Dixson & Vasey, 'Beards Augment Perceptions of Men's Age, Social Status, and Aggressiveness, but Not Attractiveness' (2012) *Behavioral Ecology*.

6   Dixson et al., 'Do Women's Preferences for Men's Facial Hair Change with Reproductive Status?' (2013) *Behavioral Ecology*.

7   Parisi et al., 'Dosimetric investigation of the solar erythemal UV radiation protection provided by beards and moustaches' (2012) Radiation Protection Dosimetry.

8   Barbeito et al., 'Microbiological Laboratory Hazard of Bearded Men' (1967) *Applied Microbiology*.

IV. WRITING

1   Oppenheimer, 'Consequences of Erudite Vernacular Utilized Irrespective of Necessity: Problems with Using Long Words Needlessly' (2006) *Applied Cognitive Psychology*.

2   Goldberg & Chemjobber, 'A Comprehensive Overview of Chemical-Free Consumer Products' (2014) *Nature Chemistry*.

3   Lander & Parkin, 'Counterexample to Euler's Conjecture on Sums of Like Powers' (1966) *Bulletin of the American Mathematical Society*.

4   Kaku, *Physics of the Future: How Science Will Shape Human Destiny and Our Daily Lives by the Year 2100* (2011).

5   Nick T, 'A Modern Smartphone or a Vintage Supercomputer: Which Is More Powerful?' (2014) *Phone Arena*.

6   'Bubble Chambers Gallery', *CERN High School Teachers*.

7   Pinckard, 'Front Seat to History: Summer Lecture Series Kicks Off' (2006) *Berkeley Lab View*.

8   Leike, 'Demonstration of the Exponential Decay Law Using Beer Froth' (2002) *European Journal of Physics*.

9   Grim, 'A Possible Role of Social Activity to Explain Differences in Publication Output among Ecologists' (2008) *Oikos*.

10 Cordell & Mccarthy, 'A Case Study of Gut Fermentation Syndrome (Auto-Brewery) with Saccharomyces Cerevisiae as the Causative Organism' (2013) *International Journal of Clinical Medicine*.

11 Soifer, Building a Bridge II: from Problems of Mathematical Olympiads to Open Problems of Mathematics (2010) *Mathematics Competitions*

12 Soifer, *How Does One Cut a Triangle?* (1990).

13 See @TwournalOf on Twitter.

14 'About', Nanopublications website.

15 Thompson et al., 'Data Publishing Using Nanopublications' (2012) *Tiny ToCS*.

16 Upper, 'The Unsuccessful Self-Treatment of a Case of "Writer's Block"' (1974) *Journal of Applied Behavior Analysis*.

17 Molloy, 'Unsuccessful Self-Treatment of a Case of "Writer's Block": A Replication' (1983) *Perceptual and Motor Skills*.

18 Hermann, 'Unsuccessful Self-Treatment of a Case of "Writer's Block": A Partial Failure to Replicate' (1984) *Perceptual and Motor Skills*.

19 Skinner et al., 'The Unsuccessful Group-Treatment of "Writer's Block"' (1985) *Perceptual and Motor Skills*.

20 Skinner & Perlini, 'The Unsuccessful Group Treatment of "Writer's Block": A Ten-Year Follow-Up' (1996) *Perceptual and Motor Skills*.

21 Didden et al., 'A Multisite Cross-Cultural Replication of Upper's (1974) Unsuccessful Self-Treatment of Writer's Block' (2007) *Journal of Applied Behavior Analysis*.

22 Mclean & Thomas, 'Unsuccsessful Treatments of "Writer's Block": A Meta-Analysis' (2014) *Psychological Reports*.

23 Moreton, 'S'More Inequality: The Neoliberal Marshmallow and the Corporate Reform of Education' (2014) *Social Text*.

24 Tsigie et al., 'A Roadmap to the Extension of the Ethiopic Writing System Standard Under Unicode and ISO-10646' (1999).

25 Reinertsen, 'Welcome to My Brain' (2013) *Qualitative Inquiry*.

26 Arnoldo et al., 'Identification of Small Molecule Inhibitors of Pseudomonas Aeruginosa Exoenzyme S Using a Yeast Phenotypic Screen' (2008) *PLOS Genetics*.

27 Garner et al., 'Reconstitution of DNA Segregation Driven by Assembly of a Prokaryotic Actin Homolog Ethan' (2010) *Science*.

28 Cole et al., 'Strong Evidence for Terrestrial Support of Zooplankton in Small Lakes Based on Stable Isotopes of Carbon, Nitrogen, and Hydrogen' (2011) *Proceedings of the National Academy of Sciences of the United States of America*.

29 Cabanac, 'Shaping the Landscape of Research in Information Systems from the Perspective of Editorial Boards: A Scientometric Study of 77 Leading

Journals' (2012) *Journal of the American Society for Information Science and Technology*, 'Unconventional Academic Writing' (2015) *Figshare*.

30   Mulligan & Sala-i-Martin, 'Transitional Dynamics in Two-Sector Models of Endogenous Growth' (1993) *Quarterly Journal of Economics*.

31   Chierichetti et al., 'Rumour Spreading and Graph Conductance' (2010) in *21st ACM-SIAM Symposium on Discrete Algorithms (SODA)*.

32   Roberts, *The History of Science Fiction* (2005).

33   'Leigh Van Valen, Evolutionary Theorist and Paleobiology Pioneer, 1935–2010' (2010) *UChicagoNews*.

34   van Valen, 'A New Evolutionary Law' (1973) *Evolutionary Theory*.

35   Thanks to John Fea for this one: Fea, 'This May Be the Best "Acknowledgments" Section of All Time' (2016) *The Way of Improvement Leads Home*.

36   Behrens et al., 'What Is the Most Interesting Part of the Brain?' (2013) *Trends in Cognitive Sciences*.

37   Marin, 'On the Residual Nilpotence of Pure Artin Groups' (2006) *Journal of Group Theory*.

38   Goupil et al., 'Rotational Splittings with CoRoT, Expected Number of Detections and Measurement Accuracy' (2006) in Fridlund et al. (eds), *ESA Special Publication*.

39   Moreton, 'S'More Inequality: The Neoliberal Marshmallow and the Corporate Reform of Education' (2014) *Social Text*.

40   Hodges, *Model Theory* (1993).

41   He & Raichle, 'The fMRI Signal, Slow Cortical Potential and Consciousness' (2009) *Trends in Cognitive Sciences*.

42   Davis, 'An External Problem for Plane Convex Curves' (1963) in *Convexity: proceedings of the Seventh Symposium in Pure Mathematics of the American Mathematical Society*.

43   'Albert Paul Malvino Quotes', *Good Reads*.

44   Brown & Henderson, 'A New Horned Dinosaur Reveals Convergent Evolution in Cranial Ornamentation in Ceratopsidae' (2015) *Current Biology*.

45   Jackson, 'Chinese Acrobatics, an Old-Time Brewery, and the "Much Needed Gap": The Life of Mathematical Reviews' (1989) *Mathematical Reviews*.

46 Reimers et al., 'Response Behaviors of Svalbard Reindeer Towards Humans and Humans Disguised as Polar Bears on Edgeøya' (2012) *Arctic, Antarctic, and Alpine Research*.

47 Nishimura et al., 'Ping-Pong Ball Avalanche at a Ski Jump' (1998) *Granular Matter*.

48 Chris Ashford, 'Bareback Sex, Queer Legal Theory and Evolving Socio-Legal Contexts' (2015) *Sexualities*

49 Neville Morley, '"The Strong Do What They Will, and the Weak Endure What They Must": Thucydidean Echoes in *Fifty Shades of Grey*' (2015) in *Thoukidideia: Occasional Publications of the Berlin Thucydides Centre*.

50 Casajus et al., 'Body Composition in Spanish Soccer Referees' (2014) *Measurement and Control*.

51 Nemiroff & Wilson, 'Searching the Internet for Evidence of Time Travelers' (2013) *arXiv*.

52 Swartz, *Why People Hate the Paperclip: Labels, Appearence, Behavior and Social Responses to User Interface Agents* (2003).

53 Minetti et al., 'Humans Running in Place on Water at Simulated Reduced Gravity' (2012) *PLOS ONE*.

54 Rubio et al., 'Characterization of Lactic Acid Bacteria Isolated from Infant Faeces as Potential Probiotic Starter Cultures for Fermented Sausages' (2014) *Food Microbiology*.

55 Turing, 'Computing Machinery and Intelligence' (1950) *Mind*.

56 Watson & Crick, 'Genetical Implications of the Structure of Deoxyribonucleic Acid' (1953) *Nature*.

57 Mendeleev, 'The Periodic Law of the Chemical Elements' (1889) *Journal of the Chemical Society*.

58 Gomberg, 'An Instance of Trivalent Carbon: Triphenylmethyl' (1900) *Journal of the American Chemical Society*. Thanks to Laura van Laeren (@lauravlaeren) for finding that one.

59 Preti et al., 'The Psychometric Discriminative Properties of the Peters et Al Delusions Inventory: A Receiver Operating Characteristic Curve Analysis' (2007) *Comprehensive Psychiatry*.

60 Riba et al., 'Subjective Effects and Tolerability of the South American Psychoactive Beverage Ayahuasca in Healthy Volunteers' (2001) *Psychopharmacology*.

61 Koski et al., 'Skeletal Development of Hand and Wrist in Finnish Children' (1961) *American Journal of Physical Anthropology*.

62 Preti et al., 'The Psychometric Discriminative Properties of the Peters et al Delusions Inventory: A Receiver Operating Characteristic Curve Analysis' (2007) *Comprehensive Psychiatry*.

63 Hainmueller & Hiscox, 'Educated Preferences: Explaining Attitudes Toward Immigration in Europe' (2015) *Sante Publique*. This phrasing appears in a preprint version of the paper, but appears to have been cut from the final published version.

64 Laulajainen, 'A Static Theory of Dry Bulk Freight Rates by Route' (2006) *Maritime Policy & Management*.

65 Clarke, 'The Transient and Steady State Responses in Oxygen Consumption by Tropical Butterflies to Temperature Step Transfer Tests' (1977) *Journal of Zoology*.

66 Glover et al., 'Heterogeneity in Physicochemical Properties Explains Differences in Silver Toxicity Amelioration by Natural Organic Matter to Daphnia Magna' (2005) *Environmental Toxicology and Chemistry / SETAC*.

67 Shcherbak, 'The Phosphorus Fractions of Muscle During Hypoxia' (1959) *Bulletin of Experimental Biology and Medicine*.

68 Unuvar et al., 'Mercury Levels in Cord Blood and Meconium of Healthy Newborns and Venous Blood of Their Mothers: Clinical, Prospective Cohort Study' (2007) *Science of the Total Environment*.

69 Wilberforce et al., 'Towards Integrated Community Mental Health Teams for Older People in England: Progress and New Insights' (2011) *International Journal of Geriatric Psychiatry*.

70 Zhu et al., 'The Effect of Emotional Conflict on Attention Allocation: An Event-Related Potential Study' (2015) *Health*.

71 Ibid.

72 Huxley, *Area, Lattice Points, and Exponential Sums* (1996).

73 Johnstone, 'On a Topological Topos' (1979) *Proceedings of the London Mathematical Society*.

74 Cox & Zucker, Intersection numbers of sections of elliptic surfaces (1979) *Inventiones Mathematicae* (named in Schwartz, 'Subgroups of Finite Index in a Free Product With Amalgamated Subgroup' (1981) *Mathematics of Computation*.)

75 Weyl, *The Classical Groups: Their Invariants and Representations* (1939).

76 Bosch, 'Unification of the Fundamental Plane and Super-Massive Black Holes Masses' (2016) *Astrophysical Journal.*

77 Bunnett & Kearley, 'Comparative Mobility of Halogens in Reactions of Dihalobenzenes with Potassium Amide in Ammonia' (1971) *The Journal of Organic Chemistry.*

78 Parent & Eskenazi, 'Toward Better Crowdsourced Transcription: Transcription of a Year of the Let's Go Bus Information System Data' (2010) *2010 IEEE Workshop on Spoken Language Technology, SLT 2010 – Proceedings.*

79 Schaul et al., 'Coherence Progress: A Measure of Interestingness Based on Fixed Compressors' (2011) *Lecture Notes in Computer Science (Including Subseries Lecture Notes in Artificial Intelligence and Lecture Notes in Bioinformatics).*

80 Cantrill, 'A Quantitative Analysis of How Often Nature Gives a Fuck' (2015) *Chemical Connectionls.*

81 Johnson, '"Art" Was a Load of Fluff' (2001) *Nature.*

82 Fairman, 'Fuck' (2007) *Cardozo Law Review.*

83 Walker, 'An Obscenity Symbol' (2015) *American Speech.*

84 Leiter, 'No Ranking is Too Trivial to Spark Commentary from Folks with Time to Burn ...' (2007).

85 Spindelman, 'Shockfreude and the Very Democratic Teachings of "Professor Fuck"' (2016) *Ohio State Law Journal.*

86 McCawley, *The Eater's Guide to Chinese Characters* (1984).

87 Dong, 'English Sentences Without Overt Grammatical Subject' (1992) in Zwicky et al. (eds) *Studies out in left field: defamatory essays presented to James D. McCawley on the occasion of his 33rd or 34th birthday.*

88 Yang et al., 'Electrochemical Synthesis of Metal and Semimetal Nanotube – Nanowire Heterojunctions and Their Electronic Transport Properties' (2007) *Chemistry Communications.*

89 Duan et al., 'Structural and Electronic Properties of Chiral Single-Wall Copper Nanotubes' (2014) *Science China Physics, Mechanics and Astronomy.*

90 Jane, '"Back to the Kitchen, Cunt": Speaking the Unspeakable about Online Misogyny' (2014) *Continuum.*

91 Dawson, 'The Compleat Motherfucker: A History of the Mother of All Dirty Words' (2009).

92   Leverette, *It's Not TV: Watching HBO in the Post-Television Era* (2009).

93   Lavery & Kim, *Reading Deadwood: A Western to Swear By* (2006).

94   Lie, *Translate This, Motherfucker! A Contrastive Study on the Subtitling of Taboo* (2013).

95   Oshiro, *Optically Enhanced Nuclear Cross Polarization in Acridine-Doped Fluorene* (1982); Cambridge Institute for Medical Research and Department of Medicine, 'Information Sheet for Research into Genetic Variation AND Altered Leucocyte Function in Health and Disease (GANDALF)' (2010); Langrish, 'Biodiesel Exhaust, Acute Vascular and Endothelial Responses' (2011); McKee et al., 'Black Intelligence Test of Cultural Homogeneity' (2011) in *Encyclopedia of Child Behavior and Development*; Zhou & Rienstra, 'High-Performance Solvent Suppression for Proton Detected Solid-State NMR' (2008) *Journal of Magnetic Resonance*; Goodacre et al., 'The RATPAC (Randomised Assessment of Treatment Using Panel Assay of Cardiac Markers) Trial: A Randomised Controlled Trial of Point-of-Care Cardiac Markers in the Emergency Department' (2011) *Health Technology Assessment*; Kivipelto et al., 'The Finnish Geriatric Intervention Study to Prevent Cognitive Impairment and Disability (FINGER): Study Design and Progress' (2013) *Alzheimer's & Dementia*; Schwitter et al., 'MR-IMPACT II: Magnetic Resonance Imaging for Myocardial Perfusion Assessment in Coronary Artery Disease Trial: Perfusion-Cardiac Magnetic Resonance vs. Single-Photon Emission Computed Tomography for the Detection of Coronary Artery Disease: A Comparative' (2013) *European Heart Journal*; Pottegård et al., 'SearCh for humourIstic and Extravagant acroNyms and Thoroughly Inappropriate Names For Important Clinical Trials (SCIENTIFIC): Qualitative and Quantitative Systematic Study' (2014) *British Medical Journal*; Ibrahim et al., 'Development, Validation, and Factorial Comparison of the McGill Self-Efficacy of Learners For Inquiry Engagement (McSELFIE) Survey in Natural Science Disciplines' (2016) *International Journal of Science Education*.

OBSCURE INTERLUDE: THE 'SCIENTIFIC' METHOD

1   Hurley et al., 'Detection of Human Blood by Immunoassay for Applications in Forensic Analysis' (2009) *Forensic Science International*.

2   Tweets by: Ethan O. Perlstein (@eperlste); NatC (@SciTriGrrl); Matthew Coxon (@mjcoxon); Alex Chase (@aechase); Myles Power (@powerm1985).

3 Bezuidenhout, 'Variations in Scientific Data Production: What Can We Learn from #OverlyHonestMethods' (2015) *Science and Engineering Ethics*.

4 Cajochen et al., 'Evidence That the Lunar Cycle Influences Human Sleep' (2013) *Current Biology*.

5 Legagneux & Ducatez, 'European Birds Adjust Their Flight Initiation Distance to Road Speed Limits' (2013) *Biology Letters*.

6 Hare et al., 'Sepsid Even-Skipped Enhancers Are Functionally Conserved in *Drosophila* Despite Lack of Sequence Conservation' (2008).

7 Bennett et al., 'Neural Correlates of Interspecies Perspective Taking in the Post-Mortem Atlantic Salmon: An Argument for Multiple Comparisons Correction' (1995). The story behind the paper is recounted in a post on *Scientific American*: Brookshire, 'IgNobel Prize in Neuroscience: The Dead Salmon Study' (2012) *Scientific American*.

V. TEACHING

1 @TheSnee and Raymond Vagell (@prancingpapio).

2 Jaschik, 'Failing the Entire Class' (2015) *Inside Higher Ed.*

3 Jaschik, 'Should Profs Leave Unruly Classes?' (2010) *Inside Higher Ed.*

4 Adapted from a real email sent to Indiana University student Tamara Millic. See Kagan, 'The Real Story About The Lover of Puppies in Party Hats' (2016) *BuzzFeed*.

5 Wainstein et al., *Investigation of Irregular Classes in the Department of African and Afro-American Studies at the University of North Carolina at Chapel Hill* (2014).

6 Waldman, 'Frontier of the Stuplime' (2015) *Slate*.

7 'Department of Transtemporal Studies' (2015) *University of Leicester website*.

8 '52-2725J Zombies in Popular Media' (2015) *Columbia College Chicago Course Catalog 2015–16*.

9 'Beckham in Degree Course' (2000) *BBC News*.

10 'Broadcasting Major (B.A.)' (2008) *Montclair State University Undergraduate Catalog 2008/2009*.

11 'What If Harry Potter Is Real?' (2011) *Appalachian State University, First Year Seminar*.

12 'DANC 127 – How Does It Feel to Dance?' (2014) *Oberlin College Course Catalog 2014-2015*.

13 Stupidity (2010) Occidental College course catalogue 2009–2010

14 Zamudio-Suaréz, 'Is Anybody Reading the Syllabus? To Find Out, Some Professors Bury Hidden Gems' (2016) *Chronicle of Higher Education*.

15 Selterman, 'Why I Give My Students a "tragedy of the Commons" Extra Credit Challenge' (2015) *Washington Post*.

16 'The Painting Reputed to Make Students Fail Exams' (2014) *BBC News*.

17 See Felton et al., 'Web-Based Student Evaluations of Professors: The Relations between Perceived Quality, Easiness and Sexiness' (2004) *Assessment & Evaluation in Higher Education*, 'Assessment & Evaluation in Higher Education Attractiveness, Easiness and Other Issues: Student Evaluations of Professors on Ratemyprofessors.com' (2015) *Assessment & Evaluation in Higher Education*. For general discussion, see Wachtel, 'Student Evaluation of College Teaching Effectiveness: A Brief Review' (1998) *Assessment & Evaluation in Higher Education*; Otto et al., 'Does Ratemyprofessor.com Really Rate My Professor?' (2008) *Assessment & Evaluation in Higher Education*; Davison & Price, 'How Do We Rate? An Evaluation of Online Student Evaluations' (2009) *Assessment & Evaluation in Higher Education*.

18 Berrett, 'In Cheeky Pushback, Colleges Razz Rate My Professors' (2014) *Chronicle of Higher Education*.

19 Boring et al., 'Student Evaluations of Teaching (Mostly) Do Not Measure Teaching Effectiveness' (2016) *ScienceOpen Research*.

20 Flaherty, 'Iowa Bill Sparks Faculty Ire' (2015) *Inside Higher Ed*.

OBSCURE INTERLUDE: FOOD, GLORIOUS FOOD

1 Zampini & Spence, 'The Role of Auditory Cues in Modulating the Perceived Crispness and Staleness of Potato Chips' (2004) *Journal of Sensory Studies*.

2 Wansink et al., 'Social and Behavioral Bottomless Bowls: Why Visual Cues of Portion Size May Influence Intake' (2005) *Obesity Research*.

3 Lan, 'Love of Instant Noodles Gets Guy into US College' (2014) *Ecns*.

4 Park et al., 'Penetration of the Oral Mucosa by Parasite-like Sperm Bags of Squid: A Case Report in a Korean Woman' (2012) *The Journal of Parasitology*.

5 Bhagat et al., 'Self-Injection With Olive Oil' (1995) *Chest*.

6   Shah et al., 'Rectal Salami' (2002) *International Journal of Clinical Practice*.

7   Humphreys et al., 'Nasal Packing with Strips of Cured Pork as Treatment for Uncontrollable Epistaxis in a Patient with Glanzmann Thrombasthenia' (2011) *Annals of Otology, Rhinology and Laryngology*.

8   Deidiker, 'Return of the Killer Fish: Accidental Choking Death on a Bluegill (Lepomis Macrochirus)' (2002) *The American Journal of Forensic Medicine and Pathology*.

9   Emsley, 'News Release: How to Make a Perfect Cup of Tea' (1987) *Royal Society of Chemistry*.

10 Burnett, *The Idiot Brain: A Neuroscientist Explains What Your Head Is Really Up To* (2016).

11 Burnett, 'How to Make Tea Correctly (according to Science): Milk First' (2014) *Guardian*.

12 Vanden-Broeck & Keller, 'Pouring Flows' (1986) *Physics of Fluids*.

13 Duez et al., 'Beating the Teapot Effect' (2009) *arXiv*.

14 Mayer & Krechetnikov, 'Walking with Coffee: Why Does It Spill?' (2012) *Physical Review E – Statistical, Nonlinear, and Soft Matter Physics*.

15 Lim et al., 'The Case of the Disappearing Teaspoons: Longitudinal Cohort Study of the Displacement of Teaspoons in an Australian Research Institute' (2005) *British Medical Journal*.

16 Conor, 'Biscuit Dunking Perfected' (1998) *Independent*.

VI. IMPACT & OUTREACH

1   For a useful overview, see: Remler, 'Are 90% of Academic Papers Really Never Cited? Reviewing the Literature on Academic Citations' (2014) *LSE Impact of Social Sciences Blog*; Barnes, 'Why Humanities Citation Statistics Are like Eskimo Words for Snow' (2016).

2   Remler, 'How Few Papers Ever Get Cited? It's Bad, But Not THAT Bad' (2014) *Social Science Space*

3   '1612 Highly Cited Researchers according to Their Google Scholar Citations Public Profiles' (2017) *Ranking Web of Universities*.

4   'Changing Perceptions of Diabetes through Stand-Up Comedy' (2015) *REF2014 Impact Case Studies*.

5   'The Management and Governance of Land to Enhance African Livelihoods' (2015) *REF2014 Impact Case Studies*.

6   Lemonick, 'Paul Erdős: The Oddball's Oddball' (1999) *Time*.

7   Goffman, 'And What Is Your Erdős Number?' (1969) *American Mathematical Monthly*.

8   Grossman, 'Facts about Erdős Numbers and the Collaboration Graph', *The Erdős Number Project*.

9   Cohen, 'Yet More Chutzpah in Assigning Erdős Numbers', *The Erdős Number Project Extended*.

10   Grossman, 'Items of Interest Related to Erdős Numbers' (2014) *The Erdős Number Project*.

11   'TIL That When Physics Professor Jack H. Hetherington Learned He Couldn't Be the Sole Author on a Paper. (Because He Used Words Like "we" & "our") Rather than Rewriting the Paper He Added His Cat as an Author' (2014) *Reddit*.

12   Cohen, 'Yet More Chutzpah in Assigning Erdös Numbers', *The Erdős Number Project Extended*.

13   Hall, 'The Kardashian Index: A Measure of Discrepant Social Media Profile for Scientists' (2014) *Genome Biology*.

14   Allen, 'We the Kardashians Are Democratizing Science' (2014) *Neuroconscience*.

15   I believe this was started by Alex Wild (@Myrmecos).

16   Jason McDermott (@BioDataGanache).

17   Nick Wan (@nickwan).

18   Jon Tennant (@Protohedgehog).

19   Blake Stacey (@blakestacey).

20   Jon Tennant (@Protohedgehog).

21   Diana Crow (@CatalyticRxn)

22   Labbé, 'Ike Antkare One of the Great Stars in the Scientific Firmament' (2010) *International Society for Scientometrics and Informetrics Newsletter*.

23   Garfield, 'The Agony and the Ecstasy: The History and Meaning of the Journal Impact Factor' (2005) *International Congress on Peer Review And Biomedical Publication*.

24   'Scientific Community Debates Standard Definition of "Impact Factor"' (2016) *C&EN Onion*.

25   Wilwhite and Fong, 'Coercive Citation in Academuc Publishing (2010) Science.

26   Davis, 'Gaming the Impact Factor Puts Journal in Time-Out' (2011) *Scholarly Kitchen*.

OBSCURE INTERLUDE: SPOOKY SCIENCE

1   Vass, 'Beyond the Grave – Understanding Human Decomposition' (2001) *Microbiology Today*.

2   Dekeirsschieter et al., 'Enhanced Characterization of the Smell of Death by Comprehensive Two-Dimensional Gas Chromatography-Time-of-Flight Mass Spectrometry (GCxGC-TOFMS)' (2012) *PLOS ONE*.

3   DeGreeff & Furton, 'Collection and Identification of Human Remains Volatiles by Non-Contact, Dynamic Airflow Sampling and SPME-GC/MS Using Various Sorbent Materials' (2011) *Analytical and Bioanalytical Chemistry*.

4   See '2012 Phenomenon', Wikipedia.

5   Wheatley-Price et al., 'The Mayan Doomsday's Effect on Survival Outcomes in Clinical Trials' (2012) *Canadian Medical Association Journal*.

6   Efthimiou & Gandhi, 'Cinema Fiction vs Physics Reality' (2006) *arXiv*.

7   Sandvik & Baerheim, 'Does Garlic Protect against Vampires? An Experimental Study' (1994) *Tidsskr Nor Laegeforen*.

8   Efthimiou & Gandhi, 'Cinema Fiction vs Physics Reality' (2006) *arXiv*. See also Mo, 'The Ethnobiology of Voodoo Zombification' (2007) *ScienceBlogs*.

VII. TWITTER

1   Priem et al., 'Prevalence and Use of Twitter among Scholars' (2011) in *Metrics 2011 Symposium on Informetric and Scientometric Research*.

2   Scoles, 'I Went to the "Contact" Radio Telescope with the Astrophysicist Behind Twitter's All-Time Sickest Burn' (2017) *Motherboard*.

3   'Academic Makes Twitter Splash Saying "Nein"' (2014) *The Local*.

4   Jarosinski, *Nein: A Manifesto* (2015).

5   Haustein & Peters, '@AcademicsSay: The Story Behind a Social-Media Experiment' (2015) *Chronicle of Higher Education*.

6   Clark, 'Twitter Joke Theft Is a Real Thing, and the Social Network Is Taking Action' (2015) *Wired*.

7   Mali, '@realpeerreview: Their Thoughts and Qualms with Academia' (2017) *Areo Magazine*.

8   Ellis, 'Sleeping Around, With, and Through Time: An Autoethnographic Rendering of a Good Night's Slumber' (2016) *Qualitative Inquiry*.

9   Nofre et al., 'Club Carib: A Geo-Ethnography of Seduction in a Lisbon Dancing Bar (2016) *Social & Cultural Geography*.

10 Rademacher & Kelly, '"I'm Here to Do Business. I'm Not Here to Play Games." Work, Consumption, and Masculinity in Storage Wars' (2016) *Journal of Communication Inquiry*.

11 Powell & Engelhardt, 'The Perilous Whiteness of Pumpkins' (2015) *GeoHumanities*.

12 Harviainen & Frank, 'Group Sex as Play: Rules and Transgression in Shared Non-Monogamy' (2016) *Games and Culture*.

13 'Episode 146: Octothorpe' (2014) *99% Invisible*.

14 Kolowich, 'Meet the 26-Year-Old Behind Academic Twitter's Most Popular Hashtags' (2016) *Chronicle of Higher Education*.

15 Lock, 'Paws for a Moment and Vote for Your Favourite Academic Cat' (2015) *The Guardian Higher Education Network*; Wright, 'Academics With Cats 2016: The Winning Photographs' (2016) *Times Higher Education*.

16 Tweets by: me; Jack Orchard (@Jarona7); Ann Loraine (@aloraine206).

17 Tweets by: Jason McDermott (@BioDataGanache); Kate Montague-Hellen (@PhDGeek); Bill Sullivan (@wjsullivan).

18 Tweets by: Michelle Booze (@DrMLBooze); Kris Weinzap (@krisilou); Viva Voce Podcast (@vivavocepodcast).

19 Tweets by: Philip Adler (@CrystPhil); Mariel Young (@Mariel_Young); Andrew Robinson (@AndrewR_physics); Mareike Ohlberg (@MareikeOhlberg).

20 Tweets by: Bilby Summerhill (@BilbySummerhill); Ameena GhaffarKucher (@AmeenaGK); Ethicist For Hire (@ethicistforhire).

21 Tweets by: Laura Howes (@L_Howes); Cafer Yavuz (@caferyavuz); Steven Vamosi (@smvamosi); John Van Hoesen (@Taconic_Musings).

22 Tweets by: me; Kelly Arbeau (@HealthPsychKell); Tracey Berg-Fulton (@BergFulton); Ben Utter(@liberapertus); Bilby Summerhill (@BilbySummerhill); me; Jeremy Noel-Tod (@jntod); Matthew Ketchum (@mattketchum).

23 Tweets by: Jon Tennant (@protohedgehog); Rebekah Rogers (@evolscientist); David Shiffman (@WhySharksMatter); Meg Fox (@NotThatMeganFox); Robinson Lab (@RobinsonLab).

24 Tweets by: Graham Steel (@McDawg); Dr Mike Whitfield (@mgwhitfield); Grouchy Grad (@GrouchyGrad).

25 Tweets by: Dieter Hochuli (@dieterhochuli); Chris Bohn (@DocBohn); Richard Emes (@rdemes); Carol Tilley (@AnUncivilPhD); Kate Maxwell (@skatemaxwell); Raul Pacheco-Vega (@raulpacheco); Erin Fisher (@DrErinFisher); Michael Kirkpatrick (@kirkpams); Laura O'Connor (@Lou_922); Lisa Munro (@llmunro).

OBSCURE INTERLUDE: LOVE AND ROMANCE

1 Russell, 'UCLA Loneliness Scale, Version 3' (1996) *Journal of Personality and Social Psychology*.

2 Harris, 'Risk of Depression Influenced by Quality of Relationships, U-M Research Says' (2013) *The University Record Online*.

3 Spielmann et al., 'Settling for Less out of Fear of Being Single' (2013) *Journal of Personality and Social Psychology*.

4 Urbani, 'Can Regular Sex Ward off Colds and Flu?' (1999) *New Scientist*.

5 Roberts, 'Scan Spots Women Faking Orgasms' (2005) BBC.

6 Innes, 'Why Falling in Love Makes You FAT: Two-Thirds of Couples Put on Two Stone after Getting into a Relationship' (2013) *Daily Mail* Online.

7 Brandspiegel et al., 'A Broken Heart' (1998) Circulation.

8 Stroebe, 'The Broken Heart Phenomenon: An Examination of the Mortality of Bereavement' (1994) *Journal of Community & Applied Social Psychology*.

9 Tweets by: Sylvain Deville (@DevilleSy); Tom Rhys Marshall (@TomRhysMarshall); Jon Tennant (@Protohedgehog).

VIII. CONFERENCES

1 See allmalepanels.tumblr.com.

2 Thanks to Shane Caldwell (@superhelical) for making these up.

3 'About Us', Academic Organization for Advancement of Strategic and International Studies (Academic OASIS) website.

4 Tweets by: Brendan A. Niemira (@Niemira); Matthew Partridge (@MCeeP); Tracy Larkhall (@TraceLarkhall); April Armstrong (@AprilCArmstrong); James Summer (@JamesBSumner); Sarah Polk (@SarahPolk); Catherine Baker (@richmondbridge); Trevor Branch (@TrevorABranch).

5 Sarah J. Young (@russianist).

6   Dan Jagger (@DrJagz).

7   Kolata, 'For Scientists, an Exploding World of Pseudo-Academia' (2013) *New York Times*.

8   Edwards, 'OMICS Group Conferences – Sham or Scam? (Either Way, Don't Go to One!)' (2013) The Cabbages of Doom.

9   It is fantastic. You can generate your own at mixosaurus.co.uk/bingo/. Credit to Kat Gupta and Heather Froehlich for the idea, and Andrew Hardie for coding.

10   Thanks to Catchclaw (@catchclaw) for that last one.

11   'Kimposium! 26 November 2015' (2016) Brunel University website.

12   Grove, 'Academia Is Keeping up with the Kardashians' (2015) *Times Higher Education*.

13   Brunel University, 'Kimposium 2015' (2015) YouTube.

14   Patton, 'How Not to Be a Jackass at Your Next Academic Conference' (2015) Vitae.

15   Ibid.

OBSCURE INTERLUDE: CAMPUS HIJINKS

1   'R2-D2 at Carleton College' (2010) YouTube.

2   Beaton, 'How Much Is Your Favorite College Football Team Worth?' (2016) *Wall Street Journal*.

3   Faulk, 'Police Beat Sept. 25–Oct. 1' (2015) *Daily Universe*.

IX.   ACADEMIC ANIMALS

1   Grossi et al., 'Walking like Dinosaurs: Chickens with Artificial Tails Provide Clues about Non-Avian Theropod Locomotion' (2014) PLOS ONE.

2   Hart et al., 'Dogs Are Sensitive to Small Variations of the Earth's Magnetic Field' (2013) *Frontiers in Zoology*.

3   Vonnegut, 'Chicken Plucking as Measure of Tornado Wind Speed' (1975) Weatherwise.

4   Dacke et al., 'Dung Beetles Use the Milky Way for Orientation' (2013) *Current Biology*.

5   Tolkamp et al., 'Are Cows More Likely to Lie down the Longer They Stand?' (2010) *Applied Animal Behaviour Science*.

6   Eichel, 'Credentialing: It May Not Be the Cat's Meow' (2011) Dreichel.com.

7   'School That Awarded MBA to Cat Sued', NBC News.

8   'List of Animals with Fraudulent Diplomas', Wikipedia.

9   Brazao, 'Big Promises, Broken Dreams' (2008) *Toronto Star.*

10   Young, 'Key EDS Witness Bought Internet Degree – Oddware' (2010) Itnews.

11   Reed & Smith, 'American University of London Sells Study-Free MBA' (2013) BBC News.

12   Myrick, 'Emotion Regulation, Procrastination, and Watching Cat Videos Online: Who Watches Internet Cats, Why, and to What Effect?' (2015) *Computers in Human Behavior*; Podhovnik, 'The Meow Factor – An Investigation of Cat Content in Today's Media' (2016) in *Arts & Humanities Conference.*

13   Fillipovic, 'Of Cats and Manuscripts' (2013) The Appendix.

14   Hetherington & Willard, 'Two-, Three-, and Four-Atom Exchange Effects in Bcc He3' (1975) *Physical Review Letters.*

15   Hetherington, 'Letter to Ms. Lubkin' (1997).

16   Willard, 'L'hélium 3 Solide: Un Antiferromagnétisme Nucléaire' (1980) *La Recherche.*

17   'APS Announces New Open Access Initiative' (2014) *APS Journals.*

18   Gartner et al., 'Personality Structure in the Domestic Cat (Felis Silvestris Catus), Scottish Wildcat (Felis Silvestris Grampia), Clouded Leopard (Neofelis Nebulosa), Snow Leopard (Panthera Uncia), and African Lion (Panthera Leo): A Comparative Study.' (2014) *Journal of Comparative Psychology.*

19   Warner, 'Demography and Movements of Free-Ranging Domestic Cats in Rural Illinois' (1985) *Journal of Wildlife Management*; Reis et al., 'How Cats Lap: Water Uptake by Felis Catus' (2010) *Science.*

20   Morell, 'Fungus Turns Frogs into Sexy Zombies' (2016) Science.

21   McAuliffe, 'How Your Cat Is Making You Crazy' (2012) *The Atlantic.*

22   Flegr et al., 'Increased Risk of Traffic Accidents in Subjects with Latent Toxoplasmosis: A Retrospective Case-Control Study' (2002) *BMC Infectious Diseases*; Flegr, 'Effects of Toxoplasma on Human Behavior'

(2007) *Schizophrenia Bulletin*; Flegr & Hodný, 'Cat Scratches, Not Bites, Are Associated with Unipolar Depression–Cross-Sectional Study' (2016) *Parasites & Vectors.*

23 Nittono et al., 'The Power of Kawaii: Viewing Cute Images Promotes a Careful Behavior and Narrows Attentional Focus' (2012) PLOS ONE.

24 Ghirlanda et al., 'Chickens Prefer Beautiful Humans' (2002) *Human Nature.*

25 Watanabe, 'Pigeons Can Discriminate "good" and "bad" Paintings by Children' (2009) *Animal Cognition.*

26 A video of the entire presentation is available online ('Chicken Chicken Chicken' (2007) Youtube).

27 Zongker, 'Chicken Chicken Chicken: Chicken Chicken' (2006) *Annals of Improbable Research.*

28 Jordanous, 'Evaluating Computational Creativity: A Standardised Procedure for Evaluating Creative Systems and Its Application' (2013).

29 Bradley, 'Crosslinguistic Perception of Pitch in Language and Music' (Prospectus) (2010).

30 Bradley, 'Teaching Portfolio' (2014) Evanbradley.net.

31 Martyniuk, *A Field Guide to Mesozoic Birds and Other Winged Dinosaurs* (2012).

32 Moeliker, 'How a Dead Duck Changed My Life' (2013) TED Talk.

33 Abrahams, 'Premiere – Homosexual Necrophiliac Duck Opera, with Scientist, at King's Cross' (2015) *Improbable Research.*

34 'Dead Duck Day', Het Natuurhistorisch website.

35 Shafik, 'Effect of Different Types of Textiles on Sexual Activity' (1993) *European Urology.*

36 Roach, Bonk: *The Curious Coupling of Sex and Science* (2009).

37 Brookshire, 'Friday Weird Science: Rats in PANTS' (2011) *Scientopia.*

38 Panksepp & Burgdorf, '50-kHz chirping (laughter?) in response to conditioned and unconditioned tickle-induced reward in rats: Effects of social housing and genetic variables' (2000) *Behavioural Brain Research.*

39 Favaro et al., 'The Vocal Repertoire of the African Penguin (Spheniscus Demersus): Structure and Function of Calls' (2014) PLOS ONE; Hospitaleche & Reguero, 'Palaeeudyptes Klekowskii, the Best-Preserved

Penguin Skeleton from the Eocene–Oligocene of Antarctica: Taxonomic and Evolutionary Remarks' (2014) *Geobios*.

40 Meyer-Rochow & Gal, 'Pressures Produced When Penguins Pooh – Calculations on Avian Defaecation' (2003) *Polar Biology*.

41 'Penguin Poo Q&A' (2005) Meyer-Rochow.com.

42 Shifman, 'ITEP Lectures in Particle Physics' (1995) arXiv.

OBSCURE INTERLUDE: MISCELLANY

1   'One of the Professors at My School. He's Always Skating around Campus' (2012) Reddit.

2   'Skateboarding Professor' (2013) Knowyourmeme.com.

3   Nitcher, 'Memetic "Skateboarding Professor" Continues Hobbies on Ground, in Air' (2012) *Daily Nebraskan*.

4   Wainwright, '"Prison-Like" Student Housing Wins Carbuncle Cup for Worst Building' (2013) *Guardian*.

5   'Yes, This Is a Student Dormitory, and Yes, There Are 1200+ Students Living in It as We Speak' (2014) Imgur.

6   'Dale Dubin: Pornography and Prison' (2007) Scrub Notes: A Blog For Med Students.

7   Massad, 'Read the Fine Print: Student Wins T-Bird' (2001) *Yale News*.

8   Vance, 'The Lives and Deaths of Academic Staplers', Tumblr.

PEER REVIEW REPORT

1   Lewis, 'Why Men Love Lingerie: Rat Study Offers Hints' (2014) *Live Science*.

2   May, 'Molecules with Silly or Unusual Names' (1997) Bristol University, School of Chemistry website.

3   Johansson & Juselius, 'Arsole Aromaticity Revisited' (2005) *Letters in Organic Chemistry*.

4   Peters & Ceci, 'Peer-Review Practices of Psychological Journals: The Fate of Published Articles, Submitted Again' (1982) *Behavioral and Brain Sciences*.

5   Piper, 'The Mysterious Disappearance of Edinburgh University's Library Cat' (2016) Stv News.

6   Delves et al., 'Item 12j: Make SUSU the Cat an Honorary President of the Students' Union (1516P25)' (2016) University of Southampton Students' Union (Open Union Council).

7   Solmi et al., 'Curiosity killed the cat: no evidence of an association between cat ownership and psychotic symptoms at ages 13 and 18 years in a UK general population cohort' (2017) *Psychological Medicine.*

8   Watson, 'Beall's List of Predatory Open Access Journals: RIP' (2017) *Nursing Open.*

9   Clancy et al., 'Survey of Academic Field Experiences (SAFE): Trainees Report Harassment and Assault' (2014) PLOS ONE.

10 David and Reed, 'Mauchly: The Computer and the Skateboard' (film, 2001)

11 Iwas A. Scientistonce, 'We Have All the Best Climates, Really, They're Great' (2017) *Journal of Alternative Facts.*

12 Hudgins, 'Hating Comic Sans Is Ableist' (2017) The Establishment.

13 Ware & Williams, 'The Dr. Fox Effect: A Study of Lecturer Effectiveness and Ratings of Instruction' (1975) *Journal of Medical Education.*

14 Merritt, 'Bias, the Brain, and Student Evaluations of Teaching' (2012) *St. John's Law Review.*

15 Matzinger and Mirkwood, 'In a fully H-2 incompatible chimera, T cells of donor origin can respond to minor histocompatibility antigens in association with either donor or host H-2 type' (1972) *The Journal of Experimental Medicine.*

16 Conner and Kitchen, *Science's Most Wanted: The Top Ten Book of Outrageous Innovators, Deadly Disasters, and Shocking Discoveries* (2002)

ANNEX I: SELECTED FIGURES

1   Andrew Franklyn-Miller et al., 'Can RSScan Footscan® D3DTM Software Predict Injury in a Military Population Following Plantar Pressure Assessment? A Prospective Cohort Study' (2014) *The Foot*; Baccante & Reid, 'Fecundity Changes in Two Exploited Walleye Populations' (1988)

North American Journal of Fisheries Management; Bleier (ed), *Comprehensive Techniques in CSF Leak Repair and Skull Base Reconstruction* (2012); Cuyno et al., 'Economic Analysis of Environmental Benefits of Integrated Pest Management: A Philippine Case Study' (2001) *Agricultural Economics*

2   Cressey, 'Grant Application Rejected over Choice of Font' (2015) *Nature*.

3   Obama, 'Securing the Future of American Health Care' (2012) *New England Journal of Medicine*, 'Presidential Policy Directive: National Preparedness' (2015) *Bulletin of the American College of Surgeons*, 'United States Health Care Reform: Progress to Date and Next Steps' (2016) *JAMA*, 'The Irreversible Momentum of Clean Energy' (2017) *Science*.

4   Paton, 'Cost of a Degree "to Rise to £26,000" after Tuition Fee Hike' (2013) *Telegraph*.

CREDITS FOR FIGURES

Figure 1: The underpant worn by the rat. Effect of different types of textiles on sexual activity. Experimental study. Shafik A.: Eur Urol. 1993;24(3):375-80. Used with kind permission of S. Karger AG, Basel.

Figure 2: Well-prepared cat. From Blanchard/Devaney/Hall. Differential Equations, 4E. © 2011 Brooks/Cole, a part of Cengage, Inc. Reproduced by permission. www.cengage.com/permissions

Figure 3: The stool collection process. An In-Depth Analysis of a Piece of Shit: Distribution of *Schistosoma mansoni* and Hookworm Eggs in Human Stool. © 2012 Krauth et al. is licensed under CC BY 2.0. Krauth SJ, Coulibaly JT, Knopp S, Traoré M, N'Goran EK, Utzinger J (2012) An In-Depth Analysis of a Piece of Shit: Distribution of *Schistosoma mansoni* and Hookworm Eggs in Human Stool. PLoS Negl Trop Dis 6(12): e1969. https://doi.org/10.1371/journal.pntd.0001969

Figure 4: Possible taphonomic scenario resulting in the accumulation of giant panda bones in the lower chamber. Remains of Holocene giant pandas from Jiangdong Mountain (Yunnan, China) and their relevance to the evolution of quaternary environments in south-western China by Nina G. Jablonski, Ji Xueping, Liu Hong, et al. in Historical Biology. Published 2017 by Taylor & Francis, reprinted by permission of the publisher Taylor & Francis Ltd, http://www.tandfonline.com).

Figure 5: Pressures produced when penguins pooh. Polar Biology, Pressures produced when penguins pooh – calculations on avian defaecation, Vol. 27, 2003, Figure 1, by Victor Benno Meyer-Rochow. With permission of Springer.

Figure 6: Cover of the first issue of Transactions. Royalsociety.org/~/media/publishing350/publishing350-exhibition-catalogue.pdf?la=en-GB

Figure 7: Strategically titled journals. Jason McDermott, RedPen/BlackPen, redpenblackpen.jasonya.com

Figure 8: Your manuscript on peer review. 'Your manuscript as submitted' by Jason McDermott, RedPen/BlackPen, redpenblackpen.jasonya.com

Figure 9: My reviews. Jason McDermott, RedPen/BlackPen, redpenblackpen.jasonya.com

Figure 10: Get me off your fucking mailing list. David Mazières and Eddie Kohler

Figure 11: The professor's lecture. *Sylvie and Bruno Concluded* by Lewis Carroll and Harry Furniss, Macmillan and Co, 1893

Figure 12: Chickens exposed to natural hair beard on mannequin. Applied and Environmental Microbiology, 1967, Vol. 14, Microbiological Laboratory Hazard of Bearded Men by Manuel S. Barbeito, Charles T. Mathews, Larry A. Taylor et al. Reproduced with permission from American Society for Microbiology.

Figure 13: The writing process. Jason McDermott, RedPen/BlackPen, redpenblackpen.jasonya.com

Figure 14: The isolator. *Science and Invention* ed. Hugo Gernsback, 1925, greatdisorder.blogspot.co.uk/2010/03/focus-focus.html

Figure 15: Academic halloween costumes. Jason McDermott, RedPen/BlackPen, redpenblackpen.jasonya.com

Figure 16: Academic valentine. Jason McDermott, RedPen/BlackPen, redpenblackpen.jasonya.com

Figure 17: Paw prints on medieval manuscript. Cat paw prints on a medieval manuscript (close up) © Emir Filipovic, 2011. Flickr.com. Reproduced by kind permission.

Figure 18: Playful, experimenter-administered, manual, somatosensory stimulation of Rattus norvegicus. Laughing Rats Are Optimistic. © 2012 Rygula et al. is licensed under CC BY 2.0, Rygula R, Pluta H, Popik P (2012) Laughing Rats Are Optimistic. PLoS ONE 7(12): e51959. https://doi.org/10.1371/journal.pone.0051959

Figure 19: Feynman diagram of bottom quark decay and 2-deminsional formula of 3,4,4,5-tetramethylcyclohexa-2,5-dien-1-one. Jason McDermott, RedPen/BlackPen, redpenblackpen.jasonya.com

INDEX

SUPPORTERS

Unbound is a new kind of publishing house. Our books are funded directly by readers. This was a very popular idea during the late eighteenth and early nineteenth centuries. Now we have revived it for the internet age. It allows authors to write the books they really want to write and readers to support the books they would most like to see published.

The names listed below are of readers who have pledged their support and made this book happen. If you'd like to join them, visit www.unbound.com.

David Adger
Denice Adkins
Paul-Michael Agapow
Christopher Aguirre
Yagiz Aksoy
António Albuquerque
Ghayda Aljuwaiser
Jaime Almansa-Sánchez
Faisal Alousi
Lisa Amir
Christian Ankerstjerne
Rita Arafa
Mark Archibald
Arekuser™ Arekuser™

arelbe
Daniel Arnesson
Maik Arnold
Richard Ashcroft
Chris Ashford
Laura E. Ayers
Emily Baker
Laurence Baldwin
Robin Ballard
Vicki Bamford
Thomas Bång
Alexandra Bardan
Jay Bardhan
Amel Barich

Amy Jane Barnes
Maria Baudin
Ruth Anne Baumgartner
Dawn Bazely
Sue Beckingham
Adrian Belcher
Ruth Bender
Regina Bento
Joanne Bernardi
Dr Joanne Beswick
Karine Betti
Tracy Bhamra
Anjlee Bhatt
Joël-Alexis Bialkiewicz

Stephen Blackham

Kim Blake

Julie Blommaert

Maureen Bode

Marie Boran

Erica Borgstrom

Richard Boulter

Simon Bradley

Katie Bridger

Laura Brimont

Robert Broad

Alice Broadribb

Julie Broken-Brow

Chris Brooke

Laura Brough

Alex Brown

Martin Brown

Lorraine Browne

Dr Christopher N. Bull

Joseph Burne

Rebecca Burton

Victoria J. Burton

Guillaume Cabanac

John Caley

Prof. I AM Canny

Tony Cantafio

Andrew Capewell

Victoria Carr

Andrew Carroll

Alice Carstairs

Anna Carter

Petra Čechová

Ondrej Cernotik

Rick Challener

David Lars Chamberlain

Emilie Champagne

Dr Justin James
    Champion

Maria Chazapis

Thom D. Chesney

Jody H. Y. Cheung

Angela Chilton

John Chrisoulakis

Vojtech Cima

Liam Clark

Jeroen Claus

Tom Cleaver

Amelia Clegg

Garrett Coakley

Stevyn Colgan

Charlotte Conradsen

Ben Cons

Jill Cooper

Sam & Mary Cooper

Debra Coplan

Jessica Corman

Ricardo Correia

Dave Cowan

John Crawford

Meriah Crawford

Tracy Creagh

Lucy Crehan

Professor Tom Crick

Ben Crosbie

Messa Da

Stephan Dahl

Michael Dalili

Albert Dang

Geoffrey Darnton

Evangelia Daskalaki

James Davenport

Andrew Davis

Jennifer Davis

Magdalena
    Derwojedowa

Lindsay DeVries

Andreas Dimopoulos

Ruth Dixon

Kevin Donnellon

Claire Donovan

Emily Dourish

Annelie Drakman

Plum Duff

Nicholas Dulvy

Robert Dumelow

David Dunbabin

Andrew Dunn

Rebecca Dunn

Vivienne Dunstan

Steven Duong

Dave Eagle

Aimee Eckert

Christian Edwards

Nick Efford

Nils Ehrenberg

Dave Ellwood

EloquentScience.com

emallson

Tomas Eriksson

Michelle Colleen Erwin

Mark Esposito

Nicola Evans

Lee Fallin

Feminist

Lori Fenton

Philip Feuerschütz

Maristella Feustle

Lenka Fiala

Deborah Fisher

Catherine Flick

Flynn Flinderson

Karina Bontes Forward

Richard Franke

Andy Franklyn-Miller

Dr. David Tyler Frazier

Steven French

John Frewin

Max Fulham

Lisandro Gaertner

John Gallaugher

Kevin Gannon

Holly Ganz

Jorge Garzon

Ivor Geoghegan

Sassan Gholiagha

Rayya Ghul

Chiara Giacomelli

James Gibson

Christopher Gilbert

Richard Gillin

Dominic Gittins

Emily Gong

Paul Goodison

Philip Gosling

Cynthia Graham

David Graham

Rebecca Gregory

Haydn Griffith-Jones

Mike Griffiths

grrlscientist

Christian Gruber

Flávia Guerra

Florence Haeneke

Crystel Hajjar

Andy Hamilton

Nicola Fox Hamilton

Youssef Hamway

Amy Harbison

Julie Hare

Sinead Harold

Lyndsey Harris

Rob Harris

Chris Hartgerink

Simon Haslam

Paul Hawkins

Kieran Healy

Dorte Henriksen

Troy Hicks

Bill Hinchen

Ralph Hippe

Tobias Hirst

Danielle Hitch

Angie Hodson

Paul Holbrook

Stephen Hoppe

Michael Hornsey

Bill Horton

Sara Houston

David Howard

Matt Huggins

Mary Hulford

James F Hutchinson

John Huxley

Sait Ilhaner

Dr H. Elyse Ireland

Lisa Jacobs

Sofiah Jamil

Meghan Jeffres

Laurent Jégou

Paul Jeorrett

Tristan John

David Johnson

Roger Jones

Abbie Jordan

Lucy Justice

jv.bg

Jamie Keen

Ursula Kelly

Luke Kemp

John Kent

Christian Strottmann
Kern

Finola Kerrigan

Dan Kieran

Steve King

Thomas Kirchner

Jens & Wencke
Kirschner

Grainne Kirwan

Yi-Juan Koh

Michelle Kothe

Nikos Kourkoumelis

Marek Kraft

Rutger Kramer

Johannes Krebs

Ivonne Kristiani

Jan Krüger

Allison Krupp

Stephan Kurz

Jakub Kuzilek

Kate Laity

Arto Lanamäki

Yann Laurans

Robert Lawson

Sally
Lawson-Cruttenden

Philippe Le Goff

Gareth Lean

Mathieu Lemaire

Marianthi Leon

Fabien Leonard

Charlie Lepretre

Shachar Lerer

Allison Leslie

Sian Lewin

Jacob Lewis

Anna Lielpetere

Mark Lilley

Gavin Lingiah

Adam Lloyd

Guilláume Lobet

Stephen Longstaffe

Angela Lord

Isabell Lorenz

Samantha Luber

Michela Luciani

Janusz Lukasiak

Cristina Lupu

Mike Lynd

Dermot Lynott

Nathan Magnall

Paul Maher

Julia Mainstone

Vladimir Makarov

Kasia Makowska

Fabian Mangahas

Kristen Mapes

Sharon Maresse

Philip Marshall

Minos Matsoukas

Kate Maxwell

Sarlae McAlpine

Jason McDermott

John A C McGowan

Claire McHale

Vanella Mead

Danielle Meder

Deepti Menon

Vlad Meşco

Ute Methner

Dagmar M. Meyer

Matt Michel

Viorel Mihalcea

Birgit Mikus

Hanry Miller

Paul Wm Miller

Margo Milne

John Mitchinson

D C Mobbs

Ali Mohammad

Luis Mojica

Ramani Moonesinghe

Annabelle Mooney

Gary Moore

Nell Morecroft

Neville Morley

Suzannah Morson

Mark Morvant

Chris Mountain

Seán Murphy

Lombe Mwambwa

Carlo Navato

Kiera Naylor

Chris Neale

Stefan Nebelung

Eamonn Newman

Jeanette Nicholas

Elizabeth Nicholson

Ivan Nikolic

Alexander Nirenberg

Elisa Nury

Kevin O'Connor

Jenny O'Gorman

Dr John O'Hagan

Kathleen O'Neill

Cheng Soon Ong

Tatiana Canales Opazo

Natalie Osborne

Jussi Paasio

Roshni Pabari

Gunnar Pálsson

Nadia Panchaud

Rosalind Parkes

Laura Pasquini

Katie Paxton-Fear

Kat Peake

Caroline Pennock

Emy Peters

Gary Phillips

Gillian Philp

Craig Pickering

Marco Piva

David Plance

Justin Pollard

Daniel Potter a.k.a.
    @legobookworm

John Powers

Lawrence Pretty

Rhian Heulwen Price

Efthymia Priki

Nick Proellochs

Manoshi Quayes

Ian Radcliffe

Jon Rainford

Rajini Rao

Colette Reap

Simon Reap

Mark Reed

Sam Reeve

Alf Rehn

Troy Gordon Rich

Cristina Rigutto

Joost Rijneveld

Laura Ritchie

Ian Rivers

Kylie Rixon

Debi Roberts

Deborah Roberts

Rachael Robinson

Jessica Robles

Julien Rochette

Stian Rødland

Wojciech Rogozinski

Daniela Rohde

Mercedes Rosello

Guido Rößling

Thomas Roulet

James Rowland

Lynne Salisbury

Jose Sallan

David Salm

AJE Sawyer

Anne Louise Schotel

Kirsty Sedgman

Claire Sedgwick

Mark Segall

Carole-Anne Sénit

Jane Shevtsov

Paulina Sidwell

Nando Sigona

Leo Simonetta

Jakob Grue Simonsen

Manmohan Singh

Erin J. Slater

Francisco Slavin

Megan Sloan

Peter Sloep

Alice Smith

Craig Smith

Mark Smithers

Tash Snaith

David Somers

Liam Spinage

Joe Spivey

Janice Staines

Sara J. Stambaugh

Kirsty Stanley

Henriette B. Stavis

Graham Steel

Len Steenkamp

Callum Stewart

Adam Stone

Casey D. Sullivan

Kimberly Swygert

Andrzej Szkuta

Javier Tabima

Elizabeth Tait

Gergely Takács

Benedict Garcia Tan

Amanda Taylor

Clare Taylor

Euan Taylor

Mark Taylor

Merja Teirikangas

Melissa Terras

Ben Thomas

Kate Thomson

Jon Thrower

Alejandro Uribe Tirado

Blanca Torres-Olave

Youcef Toumi

Ashley Towers

Mark Townley

Julia Tratt

Andrea Travaglia

Alison Traweek

Donald J. Trump

Bradley Turner

Lucina Uddin

Raymond Vagell

Jace Valcore

Tim van der Zee

Jan Verstraete

Vetenskap och Folkbildning Skåne

Alice Violett

Jose Vizcaino

Viola Voß

Matt Waite

David Wakefield

Simon Waldman

Brian Wansink

Brittany Warman

Bart Wasiak

Gotthard Weiss

Julius Welby

Annie West

Chris Weston

Margaret White

Kyla Whitefoot

Douglas Whiteside

Mike Whitfield

Leila Whitworth

F.D.C. Willard

Gareth Williams

Sean Williamson

Derek Wilson

Isabelle Winder

Kurt Winkelmann

Maria Wolters

Iain Wood

Imogen Lesser Woods

Michelle Worthington

Colin & Rachel Wright

Dawn Wright

Ian Wright

Ken Wright

Yoriko Yamamura

Jean & Paul Yates, Ryan, Ben & Dan

Hank Yeomans

Cameron Yick

Linda Youdelis

Damon Young

Peter Young

Carl-Mikael Zetterling

Sáni Zou